Writing Public Policy

❖

A Practical Guide to Communicating in the Policy-Making Process

Catherine F. Smith

New York   Oxford
Oxford University Press
2005

Oxford University Press

Oxford   New York
Auckland Bangkok Buenos Aires Cape Town Chennai
Dar es Salaam Delhi Hong Kong Istanbul Karachi Kolkata
Kuala Lumpur Madrid Melbourne Mexico City Mumbai Nairobi
São Paulo Shanghai Taipei Tokyo Toronto

Published by Oxford University Press, Inc.
198 Madison Avenue, New York, New York 10016
www.oup.com

Oxford is a registered trademark of Oxford University Press

Library of Congress Cataloging-in-Publication Data

Smith, Catherine F. (Catherine Findley), 1942-
    Writing public policy: a practical guide to communicating in the
    policy-making process
Catherine F. Smith.
        p. cm.
    Includes index.
    ISBN-13: 978-0-19-514507-6
    ISBN 0-19-514507-0 (pbk.: alk. paper)
    1. Communication in public administration.   2. Written communication.
I. Title.

JF1525.C59S64 2005
320.6 – dc22

                                                          2004056062

Printing number: 9 8 7 6 5 4 3 2 1

Printed in the United States of America
on acid-free paper

❖ Contents ❖

❖ Preface ❖

A student returning to campus from a summer internship in a Washington, D.C., public policy think tank had this to say about a lesson she learned from the experience: "In public policy work, if you can't write it or say it, you can't do it."

My experience as a communications consultant to government tells me that she is essentially correct. As a teacher of writing, I know that communication skill combined with know-how can make a difference. I wrote this book to prepare students and others to effect real change by writing (and talking) to "do" public policy in democracy.

What Is the Purpose of the Book?

It is a practical guide to writing and speaking during public policy making processes. It aims to develop communication know-how and skill. Know-how means knowing what to do or having the ability to interpret situations in context. Skill means knowing how to do, or having competencies ready to use.

It does:

- Describe the public policy making process
- Identify communication's functions and limitations in that process
- Explain standards and expectations for communicating in the public sector or between the public and private sectors
- Guide the use of selected public policy communication genres

It does not:

- Discuss theory of public policy, writing, or communication
- Teach introductory public policy analysis, written composition, public speaking, or public communication

What Is the Book About?

Public policy making in a democracy is an institutional process of solving problems that affect human society or its environment. Communication—both written and oral—is integral to the process.
A public policy process includes the following activities:

- Defining the problem
- Developing knowledge of prior action or inaction on the problem
- Proposing policy alternatives
- Deliberating the alternatives
- Adopting policy
- Administering and implementing policy
- Changing policy

Actual processes are not so linear as this list suggests. The list should not be taken as a step-by-step description of what happens in every instance. Rather, it represents generic activities in a logical order. Real public policy making incorporates those activities, but they do not always occur in this order, or in a single pass, or in a simple way. Actual processes are messy.

Each activity relies on associated communication practices and their products. For example, defining the problem might generate a set of notes for personal use on experiences with problematic conditions, an oral briefing for an elected official about the problematic conditions, a memo for a legislative committee chair summarizing analysis of policy options—all of those and more. Problem definition is conducted by composing, presenting, and exchanging such communications.

Who Is the Intended Audience?

Primarily, the book addresses undergraduate or graduate students of public policy or political science or communication or writing. As a practical, concise guide to writing and communicating in the public policy process, it is intended for courses such as "Public Policy Analysis," "Public Administration and Policy," "Issues in Public and Professional Discourse," "Writing for the Public Interest," "Rhetoric and Public Affairs," "Civic Writing," or "Writing in a Democratic Society." It can be used in advanced writing courses, professional communication courses, or in any writing-intensive course that deals with societal or environmental problems or aspects of the public policy process.

Another intended audience for this book is entry-level professionals in fields such as management, politics, government, public relations, law, public policy, journalism, social work, or public health. This text will be a useful tool to any professional in a role concerned with public affairs who seeks to improve his or her writing and communication skills.

This book will also appeal to any active citizen with an interest in a community issue, politics, or government. *Writing Public Policy: A Practical Guide to Communicating in the Policy-Making Process* will help the active citizen to bring about change by providing an understanding of democratic public processes and the skills needed to perform conventional types of communication commonly used in solving public problems.

Acknowledgments

This book results directly from my interaction with professionals and executives in federal and state government, with elected and appointed officials in local government, and with students in the Public Affairs Program of the Maxwell School of Citizenship and Public Affairs at Syracuse University as well as students in the Technical and Professional Communication Program of the English Department at East Carolina University. I especially acknowledge Kathy Karlson of the U.S. Government Accountability Office (GAO) Training Institute and my seminar co-leader, JoAnn Crandall, formerly of the Center for Applied Linguistics, for the book's original impetus. The exemplary know-how and skill of Nancy Kingsbury, Kenneth Mead, Keith Fultz and others (too many to list) at GAO set the book's goals. I especially thank Syracuse University students Nicholas Alexander, Randy Ali, Sandra Derstine, Felicia Feinerman, Elizabeth Graves, Alaina Miller, Lisa M. Mueller, and Jennifer Salomon, as well as University of North Carolina student Heidi R. Karp for permitting this book's uses of their policy-making experiences and for generously contributing sample documents. Similarly, I thank (former) Orange County (North Carolina) Commissioner Margaret W. Brown, Gregg Township (Pennsylvania) supervisor Douglas P. Bierly, Penns Valley (Pennsylvania) Conservation Association board members, and (former) Policy Director of the U.S.

Department of Transportation's National Highway and Traffic Safety Administration Carl E. Nash.

While the book was in progress, many academic colleagues and community friends provided materials, advice, or support. Syracuse University's Vision Fund enabled the piloting, and the College of Arts and Sciences supported the development, of an undergraduate course in writing public policy from which the plan for this book derived. John B. Smith and Carl E. Nash encouraged and helped the project in countless ways. For contributions to individual chapters, I gratefully acknowledge Syracuse University colleagues William D. Coplin, Carol Dwyer, and Frank Lazarski of the Public Affairs Program; Natasha Cooper and Lesley Pease of the E. S. Bird Library; Stephen Thorley, Frederick Gale, and Molly Voorheis of the Writing Program; and Krista Perreira, Daniel Gitterman, and Mort Webster in the University of North Carolina–Chapel Hill's Department of Public Policy. Special thanks to Stephen Thorley and Molly Voorheis for critiquing the book from the professional practitioner's and writing teacher's perspectives. Special thanks to Susan Warner-Mills, Jinx Crouch, and Suzanne D. LaLonde for critiquing it from the active citizen's perspective. Special thanks to George Rhinehart, Heather Rosso, and Alex Bierly for expert manuscript preparation.

Reviewers' detailed suggestions improved the book's usefulness. I gratefully acknowledge Robert V. Bartlett, Purdue University; Charles Bazerman, University of California, Santa Barbara; Marie Danziger, Harvard University; Amy Devitt, University of Kansas; James Dubinsky, Virginia Tech; Daniel Gitterman, University of North Carolina at Chapel Hill; David Kaufer, Carnegie Mellon; Joe Olson, Pennsylvania State University; Carolyn Rude, Virginia Tech; and Susan Tolchin, George Mason University. Oxford University Press acquisitions editor Tony English, development editor Jan Beatty, associate editor Talia Krohn, project editor Celeste Alexander, and copyeditor Sara Carrier made the book possible and made it better.

Finally, I dedicate the book to my family: John (best friend and best critic, who serves on a municipal water and sewer authority); Ian (who serves on a municipal planning commission); Lauren (who administers a statewide nonprofit organization); Jimmy (who serves on city and county utilities commissions); Floride (who managed county voter registration and elections); Marian, David, all the children, and in memory of Helen and Jim.

❖ Introduction: How to Use This Book ❖

To benefit from this book, you do not need prior experience as a student of public policy, as an intern in government, as a student in community-based learning, as an activist in campus or community affairs, or as a volunteer in a nonprofit organization. Any such experience will be very helpful, and you can draw on it often as you use the book. But it is not required. Experience or training in professional or business or administrative communication will be helpful too, but it is not required.

You will find this book useful if:

- You are majoring in the social sciences or humanities to prepare for a career in politics, government, public relations, law, public policy, journalism, social work, or public health
- You are (or might in the future be) an intern in government, in a think tank, or in a nongovernmental organization concerned with public affairs
- You are preparing to enter (or already practicing in) a publicly regulated industry or business
- You have (or seek) a job as a communications aide in government or a political action organization
- You have (or seek) a job as a public policy/public relations director in a nonprofit organization or as a public affairs liaison in a corporation, trade organization, professional association, or community service agency
- You are a writer, and you write about public affairs
- You are concerned about a local, national, or world problem, and you want to do something about it

- You want or are asked to comment publicly on a controversy, and you do not feel you have the authority or knowledge or skill to do that

How Should You Use the Book?

It is a manual of practice, so you can use it whenever you have a practical need to enter a public forum or to participate in a public process of making policy. Here's the intended model of use. You bring experience, knowledge, and willingness to learn. Additionally, you supply a topic of concern (or an assignment does). Review the foundational chapters 1 and 2. Then follow the instructions (chapters 3 through 9) to write a needed document or prepare a needed talk. Before you write, answer the general method's questions (chapter 2) to consider the policy context and rhetorical situation and to plan the communication. After you write, use the checklists of expected qualities (chapter 2) to assess the product.

Do you need all of the book or only parts? The chapters build on each other, but they can be used separately. If you are using this guide as a textbook in a writing or communications course, you might use it all. Start with substance, or the topic of concern; then compose the sequence of communications in order to "work" the topic through a typical policy process. Work straight through, from defining the problem to commenting on the administration of policy. You might do this over the duration of the course. That way, you increase your knowledge of the topic, you gain an overview of policy-making process, and you practice a set of integrated communication skills. For a collaborative project to be done in a shorter time, you might divide up the communication responsibilities. That way, each member gains practical experience with one or more of the types of practical communication and the group gains an overview.

If you are learning independently, use this guide chapter by chapter, as described. If you are teaching others, use the chapter sequence as a skeleton to be fleshed out by assignments.

Incidentally, if at the outset you feel intimidated because you don't know enough about a topic or about policy making, do not worry. Your knowledge of substance and of process will grow as you practice each type of communication. Starting only with a topic of concern, you will know much more about the topic after you conduct the inquiry or analysis needed to define it as a policy problem (chapter 3).

You will learn about the policy process by searching public records for prior government action on the problem (chapter 4). At that point, you will probably feel overwhelmed by the amount of available information. However, you will start to filter for relevant information as you clarify your own position on solving the problem that interests you (chapter 5). And, having learned something about existing policy, you are beginning to formulate an alternative (chapter 6). When opportunity comes or circumstances demand, you'll feel ready to articulate your viewpoint in a public process (chapters 7, 8, 9).

If you have a singular assignment or a specific need, use only the parts of the book that you need. In any case, you should read (or skim) chapters 1 on public policy and 2 on communication in the policy-making process. Then select and use the chapter(s) that will help you to produce a needed communication. For example, if you are using the book in a public policy analysis course, you might read chapters 1 and 2 and then (depending on the assignment) select chapter 3 on defining policy problems or chapter 7 on informing policy makers. If you are working, interning, or volunteering in an organization and have an obligation or an opportunity to intervene in a policy-making process, after reviewing chapters 1 and 2 you might choose chapter 8 on oral testimony or chapter 9 on written comment. (Here's a heads-up: Several chapters anticipate responsibilities that are frequently given to interns, volunteers, and entry-level professionals in public policy workplaces, ready or not. They are chapter 3 on problem definition, chapter 4 on legislative records research, chapter 8 on witness testimony, and chapter 9 on written public comment.)

How Should You Read the Book?

Whether you are using the book as a textbook or a reference, knowing what to expect can be helpful. Here is a summary of its contents and organization.

To develop know-how and skill, this guide incorporates several different learning strategies:

- Scenarios to make the policy process real and to illustrate the diversity of contexts, situations, and actors involved
- Sample real-life documents
- Commentary on the samples connected to a general communication method and to specific instructions

- General method for planning, producing, and assessing policy communications, informed by experience in democratic public institutions
- Specific instructions for composing and presenting written or spoken communications that are commonly used in public policy processes

These learning strategies reflect real-world conditions. Every scenario and sample is real, taken from actual or simulated policy processes. Samples illustrate typical communication products; many were written by communicators shown in the scenarios. The samples are just that, actual samples, not models. (The samples are presented as written. In some, titles have been supplied or deletions noted where details have been omitted due to length.) The scenarios and samples provide a realistic and multifaceted overview of the process and its products while not oversimplifying them. This guide's "immersion" approach to developing know-how assumes that broad exposure creates confidence by developing a sense of predictability or a feel for the territory. Consequently, you will not find here a singular case of policy making to which all scenarios and samples correspond. Instead, you will find a wealth of exemplification drawn from multiple policy processes. Coherence is provided by selected topics (health and safety, for instance) that persist like themes in a novel. Public health and safety show up often in this book, as they do in real-life policy making. Here, they thematically underline the ordinary complexity of policy making as problems persist and diverse actors engage them for different purposes in changing political climates.

Given the real-life circumstances of policy making, this guide's approach to skill development assumes that public policy communicators need both general and specific skills. General ability to consider contexts and situations rhetorically, strategically, and ethically is important. General ability to judge a communication's potential impact is important, too. Specific abilities to compose expected types of communication are obviously necessary. Consequently, a general method of rhetorical and strategic planning is provided here, along with detailed task-by-task instructions for commonly used genres or types of communication. Together, the method and instructions provide a disciplined approach to writing and speaking successfully under typical working conditions of public policy activity.

Three limits of this guide should be noted. First, the emphasis is on communicating in, or with, government. The emphasis is justified both because government makes most public policy and because many communicators know too little about what government does and does not do. While it is true that large-scale corporate or religious or educational institutions also make public policy, those actors are addressed here only as influences on governmental policy making. Second, all scenarios are drawn from U.S. federal, state, or local government because those sources are more accessible than other national sources. Third, some less familiar types of institutional policy communication are included here while more familiar types of mass public communication are excluded. For example, oral testimony in legislative hearings and written comment in administrative rulemaking are included, but press releases and letters to the editor are not. Familiar communications are only included if help with them cannot easily be found elsewhere. For instance, to write a press release, you can get help in academic public relations and professional communications courses. Excellent practical advice is also provided in Reeher and Mariani's *The Insider's Guide to Political Internships* (185–94). For public letter writing, you can get help in civic writing courses or social activism workshops. Also, websites of national civic organizations and nonprofit advocacy groups frequently offer guidance both on writing letters and on producing press kits.

With this introduction, I hope you feel ready to look further into the guidance offered here. The next two chapters give you background on the policy-making process and communication within it. In the following chapters, you'll find the instructions for specific types of written or oral presentation. May you enjoy the activity and communicate well to do good work.

Reference

Reeher, Grant, and Mack Mariani. *The Insider's Guide to Political Internships.* Boulder, CO: Westview Press, 2002.

❖ 1 ❖

PUBLIC POLICY MAKING

Public policy exists to solve problems affecting people in society (Coplin and O'Leary 3). Making public policy means deciding what is and is not a problem, choosing which problems to solve, and deciding how to solve them to benefit society. People are likely to differ on the problem and on how to solve it, so conflict is to be expected. Typically, governments decide which problems will be addressed and how to address them. A frequently quoted definition says that public policy is whatever governments choose to do or not do (Dye 1). Other definitions emphasize that government is only one policy-making institution; cultural authorities in religion or business or education make public policy, too. Still other definitions shift the focus from authority to the activity of making policy choices. Those definitions widen the scope of choice-making beyond political institutions, emphasize the importance of process, and point to the influence of context on decision (Clemons and McBeth; MacRae and Whittington). From the perspective of functional communication, the most important characteristic of policy-making activity is that it is a process occurring in a context.

As problem-solving activity, public policy making has three basic components: the problem, the policy, and players. A problem is something perceived to be wrong in a society or its environment. A policy is a goal with a plan of action to solve the problem. A player is an influential participant in the process (Coplin and O'Leary 8). Problems, policies, and players function within political systems and

public discourses to give those components their real meaning in particular contexts and situations.

To illustrate, in the United States, health care costs are rising too
fast and too many people are unable to afford health care. That's a
problem. In general public discourse, most people agree that it is a
problem, although there is disagreement on the causes and effects.
Most people agree that something must be done, but there is disagreement on what to do. That's democracy. In 1993, the federal
government acted. The Clinton administration proposed to guaran-
tee health care for all Americans from birth to death. The proposal
had two goals, expanding health care coverage to uninsured people
and reducing the rate of spending on health care. The plan of action included consumers being organized into alliances to bargain
for cheaper health care; employers and employees sharing costs of
health insurance; care providers offering a basic package of services
for guaranteed fees; government fixing caps on fees to reduce costs;
and federal administrators setting standards for care, protecting consumers' rights, and ensuring that providers fulfilled their responsibilities. That's a *policy*. Members of Congress proposed alternative
policies. Lobbyists for a spectrum of interested groups informed and
influenced all the proposals. The Clinton administration advocated
their proposal. The development of these proposals represented
institutional (governmental) discourse about a particular problem.
Influential participants who made proposals — the president, legislators, and lobbying groups — were the *players*. In the end, none of the
proposals was adopted. Instead, legislators modified the existing
health care system to incorporate some elements of the administration's plan and some elements of alternative plans. That's *politics*.
The outcome is the present policy solution to the public problem of
health care. (The description is taken, with added interpretation,
from Wolpe and Levine's *Lobbying Congress: How the System Works*,
2nd ed. [104–116].)

An overhaul of the national health care system is an extraordinary
initiative. For understanding public policy making as an ongoing
democratic process, ordinary or "unsexy" (as journalists like to say)
illustrations might be better. Budgeting exemplifies these regular
workaday processes. An actual state budget development is described
below, shown from the viewpoint of the communications director for
the chairman of a state senate's budget committee.

Preview. The annual state budgeting process occurs over six months with preset deadlines or milestones. In January, the governor proposes a budget for the coming year that represents the administration's priorities and politics. The legislative committees respond in March (for the house) and in May (for the senate) with recommendations based on their priorities and politics. Effectively, three budget proposals—the governor's, the house's, and the senate's—must culminate in a single adopted budget by July 1, the mandated start of the state's new fiscal year.

Scenario

In early January, a state governor holds a press conference to announce the release of his proposed budget for the coming year. Immediately after the governor's press conference, the chairs of the state's house of representatives and senate budget committees comment publicly on the governor's proposed budget in other press conferences, newspaper interviews, and radio and television talk show appearances. The communications director for the senate budget chair tracks public response to the governor's budget and to the senate chair's comments on it.

At the same time, work begins on the senate and house budget recommendations. In the senate, the current chair of the ways and means committee brings his staff (an administrative assistant and Steve, the communications director) to a meeting with staff for the permanent committee. Present are the ways and means chief of staff, chief legal counsel, and chief budget analyst. The chair has authority, as a member of the majority political party, to set the senate's current budget policy. The permanent committee staff has responsibility for developing, with the help of the chair's staff, the senate's recommendations for budgeting according to current priorities.

In the first meeting in January, the chair and the combined staffs review budget history (what's carried over from last year and what's new this year), the state of the economy (current and projected conditions), and the politics of individual budget items (item is nice to have but can be sacrificed if necessary, item is nonnegotiable, we expect a fight on the item, or we go to the mat with the item). They compile a rough list of potential priorities for the coming year's budget. Because he will draft text for the senate recommendations, the communications director starts taking notes.

After the first meeting, the committee staff fans out in January and February to consult with federal and state fiscal experts, as well as with experts on specific issues in state agencies, government watchdog groups, and advocacy groups. They get more projections for the economy, and they seek external corroboration for their rough list of budget priorities. The communications director goes along to all these consultations.

Next, the committee staff solicits budget requests internally from senate members, state departments, and state agencies. Staffers meet with the members, departments, and agencies about their requests. They begin an initial breakdown of line items to include in the senate recommendations. The communications director stays in touch with the staff. In parallel, he maintains daily or weekly contact with editors and reporters of major news media. He develops relationships and educates the press. They, in turn, keep him up-to-date on budget-relevant news. He maintains good contact both internally and externally because he has dual responsibilities to anticipate debate about the senate's recommendations and to present them in a way that will promote their acceptance by government officials and the public.

A second working meeting is held. The chair and combined staffs intensely debate priorities and preliminarily decide on key priorities. Later in March, when the house budget proposal is released, the combined staffs analyze it, compare it to the governor's proposal, and compare it to their own developing proposal. The communications director participates in the meetings and continues to track press and public responses to the governor's proposal and to the house recommendations. Most important, he translates the key priorities (decided at the second working meeting) into key messages, simple statements that identify a key issue and the senate's proposed way of using tax dollars to address the issue. He gets the chair's and committee senior staff's commitment to emphasize the key messages at every communication opportunity. Whenever they speak or write, they agree that the key messages will be appropriately included.

Throughout March and April, the senate budget committee staff finalizes its recommendations and interacts with the governor's and house committee's staffs. The communications director's attention increasingly turns to his primary responsibility of drafting the document that will both present senate recommendations and publicize them; he must also prepare for debate in the legislature and for negotiation with the governor's office during the budget approval process.

In March, he writes preliminary drafts of the chairman's introduction and the executive summary for the document. (He knows that when the lengthy and detailed document is released, many people, including the press, will read only the chair's introduction and the executive summary.) He emphasizes the key messages in both. He writes (or edits senior staff's) descriptions of major budget categories (health care, education, housing, and so forth). From his notes taken in budget working meetings, he develops arguments to support proposed dollar figures for existing line items and new initiatives in each category.

Also in March, he plans a comprehensive internal and external presentation strategy to be carried out in June. Along with internal distribution to the governor, the legislature, and government departments and agencies, the senate's recommendations will be publicized through an external news media and public events campaign conducted before, during, and after formal release of the recommendations document.

In April and early May, he revises the document based on committee staffers' review of his preliminary drafts and edits of their drafts. He coordinates with news media and advocacy groups regarding a public relations campaign to accompany release of the senate recommendations. By mid-May, the finished 600-page document presenting the recommendations is delivered to the printer. He fields inquiries by the press and the public about the soon-to-be-released recommendations, and he focuses on writing, editing, and revising press releases, other public announcements, and the chairman's comments for the senate budget release press conference.

In late May, the senate recommendations are released, distributed, and announced. Simultaneously, the planned public relations campaign is conducted. Throughout June, while the senate and house debate the budget and the governor responds to their debates, events all around the state (preplanned jointly by the communications director and advocacy groups) direct public attention to senate priorities and funding proposals dur-

ing "health care week" or "education week" or "citizenship assistance week." Meanwhile, back in the senate, the communications director puts out daily press releases, follows up phone
contacts by the press or the public, and prepares comments for
the chair's use in responding to unexpected developments, politically significant news, or budget controversies.

What This Scenario Shows

This detailed scenario shows public policy making in process. The
problem is the need to finance state government operations and public services in the coming year. The process is the annual budgeting
cycle. The major *players* are three elected officials (the governor and
the chairs of the state senate and house of representatives budget
committees). Five appointed professional staffs (the governor's, the
two chairs', and the two committees') advise and assist the elected officials. Other players are experts inside and outside state government
with knowledge on specific topics, policy analysts who will advise authorities on ways to approach the problem, and advocates representing special civic, commercial, or political interests in the solution.
The resulting *policy* is a set of priorities and related recommendations
for spending.

From this scenario, you might be able to see components of policy
making functioning in a flow of actions to conduct a process. In
budgeting, basic institutional actions are these: to define priorities in
relation to current conditions and goals; to review previous goals; to
take reasoned positions on needs, argue for them, and negotiate with
others who reason differently; to propose specific objectives and
spending levels; to deliberate alternative proposals and decide; and
to inform and invite public participation. The flow of activity in this
particular process is typical of institutional policy making.

Typical integration of communication and action is shown here,
too. Communication products materialize the action and enable further action. For example, what most people call "the budget" is not
the policy itself but rather an intentionally persuasive document
(composed by the communications director, in this instance) that argues for objectives based on the priorities and that proposes funding
allocations to accomplish them. It is only the last of many documents
that move the process along. At earlier stages, working discussions
are materialized in draft documents. Circulation of the drafts for

comment, editing, and revision facilitates negotiation about priorities. With persuasive expression and specific figures, the final budget document serves both general public discourse (persuasive expression of priorities and objectives) and institutional discourse (specific figures) about governmental spending in the coming year.

Practical aspects of communication in this scenario deserve comment, too. From a communicator's viewpoint, the budgeting scenario, while orderly as an annual and scheduled process, is quite messy in reality. The scenario suggests the density of information, variety of demands, balancing of competing interests, coordination of roles, even the juggling of schedules that characterize a policy process and create the working conditions under which communications are produced. For communicators, this scenario shows well the need for a disciplined approach that keeps you on track, enables you to produce under pressure, and supports accountability in the process.

Such an approach is presented in the next chapter. First, however, the creation and use of information in public policy making is explained.

References

Clemons, Randall S., and Mark K. McBeth. *Public Policy Praxis*. Upper Saddle River, NJ: Prentice Hall, 2001.

Coplin, William D., and Michael K. O'Leary. *Public Policy Skills*. 3rd ed. Washington, D.C.: Policy Studies Associates, 1998.

Dye, Thomas J. *Understanding Public Policy*. Englewood Cliffs, NJ: Prentice Hall, 1987.

MacRae, Duncan, Jr., and Dale Whittington. *Expert Advice for Policy Choice*. Washington, D.C.: Georgetown University Press, 1997.

Wolpe, Bruce C., and Bertram J. Levine. *Lobbying Congress: How the System Works*. 2nd ed. Washington, D.C.: Congressional Quarterly Inc., 1996.

❖ 2 ❖

COMMUNICATION
IN THE PROCESS

"In public policy work, if you can't write it or speak it, you can't do it."

The undergraduate who said that (after experience in a public policy internship) is right on target. Writing and speaking are not sufficient to make public policy, but they are necessary. Communication enables the process in two fundamental ways.

1. Communication produces useful information. Useful information in a public policy process has four major characteristics: it helps to solve problems, it is action-oriented, it has consequences, and it is publicly accessible.

- **Helps to solve problems.** Information is needed at every stage of a policy process — to frame problems, to analyze issues, to debate approaches, to find and decide on solutions. Only relevant information is useful, however. In deciding whether to provide information, always ask and answer these questions: How will it help to solve the problem? Whom will it help?
- **Is action-oriented.** In policy work, information makes things happen. In deciding whether and how to inform in a policy process, always ask and answer these questions: What do I want this information to do? What effect might this information have?

- **Has consequences.** A problem and its solution affect other problems and solutions in many contexts. Consequently, a policy's effects can be wide-ranging. In deciding whether and how to inform in a policy process, always ask and answer these questions: What is likely to happen as a result of this information? What impacts might this information have?
- **Is publicly accessible.** Policy makers are answerable to the people who give them authority. Therefore, information used in public processes must be publicly available. Officially, it is recorded and preserved by government as an authoritative public record. Unofficially, news media and people in everyday social interactions distribute information as well. In deciding whether and how to inform a policy process, always ask and answer this question: How will this information be made public?

2. Communication makes information intelligible in context. As it is meant here, context is, narrowly, the public policy process for which information is produced. (The wider meaning of context as anything that might influence a communication is a bit too broad for the purposes of this practical guide to communicating in policy-making contexts.) Intelligibility is the two-way transaction by which communicators use shared knowledge of expectations to create and interpret useful information.

To make information intelligible in context, writers (and speakers) must ensure that recipients can recognize the type of communication underway. Writers (and speakers) do this by knowing, themselves, typical purposes (for instance, problem definition) and the range of document or speech types conventionally used for the purpose. Sometimes called genre knowledge, this kind of know-how involves general ability to understand forms of communication in relation to their function in a context and their effect on the context. To develop that kind of know-how in order to enable appropriate choices by writers (and speakers) is an overall aim of this guide.

Writers (and speakers) and their recipients must share knowledge of expected standards if a document or talk is to accomplish its objective.

Standards

In public policy communication, what matters most is not how much you know but rather how much your readers or listeners know after they have read your writing or heard you speak. Information is expected to be *useful*.

Presentation is expected to be *clear, concise, correct, and credible*. Public policy work is information-overloaded. Especially in government settings, time is scarce, schedules are nearly impossible, and attention is always divided. Rarely does anybody have patience for disorganized, wordy communication or information that does not serve a purpose. Information functions best when it can be comprehended quickly, trusted as accurate, traced to authoritative sources, and used with confidence.

In the Preface to this book, it is claimed that policy actions are associated with communication practices. Following are some common practices associated with the actors who use them to perform functional roles in a policy process.

Participants, Roles, and Practices

Who generates public policy information? Actors in a policy process do. Typically, actors are the players, varied professionals inside and outside government, and advocates.

Players.   Players represent organized interests in a policy process. Unorganized interests—for example, individuals affected by a problem or a policy—are not players, typically. While individuals acting alone can sometimes make a difference in a process, they are not usually players. To be a player in a policy process requires collective and organized effort (Coplin and O'Leary 7).

Typical players in public policy processes include:

- Providers of goods, services, or activities related to the problem
- Consumers of goods or services in the problem area (if organized)
- Experts with specialized knowledge of the problem
- Advocates and lobbyists representing particular interests in the problem
- Officials with authority to solve the problem

For example, in making policy for highway safety, the following players would be involved:

- Automotive and insurance industries as providers of goods, services, or activities
- Organizations of automobile drivers as consumers
- Specialists in automobile design or analysts of the economics of transportation as experts
- Advocates for accident victims and lobbyists for law enforcement associations as representatives of particular interests
- Members of Congress, cabinet secretaries, or state governors as official authorities

Whether they write or speak themselves or they authorize others to do it for them, players generate information in relation to their role or for which they are the credible source. In the auto safety example, automotive industries might communicate technical information on safety features of vehicles. Insurance industries might communicate information on the economic consequences of accidents. Consumer groups might offer accounts of experiences in using automotive products and identify problematic conditions. Expert specialists in automobile design or materials might offer results of research on ways to make cars safer. Expert analysts might offer advice regarding policy choice such as regulation of manufacturers versus education of consumers. Advocates and lobbyists might provide germane information about interested or affected groups, propose policy, and argue for or against policy based on group interests. Elected and appointed officials generate the policy instruments (for example, reallocate funds, create a new program, or provide more oversight for existing programs).

Roles.   Players' role-related communications are presented in forms conventionally used for the specific purpose. For example, elected and appointed legislative officials use bills and resolutions. Administrative officials use executive orders, statutes, legal codifications, standards and rules of enforcement, and programs of implementation. (You can learn more about these document types in chapter 4 on government records research, where you are referred to respected sources such as the Library of Congress's database Thomas that includes glossaries of legislative and executive documents.) Advocates use position papers, research reports, and press releases.

Professionals Inside Government.   Within government, diverse career or consulting professionals generate most of the working information of a policy process. They communicate in roles as, for example, legislative aides to members of a legislature; experts on the staffs of legislative committees; legal counsels to legislative committees and agencies; executive agency administrators; policy analysts and technical specialists attached to many offices. To carry out their responsibilities, they might use any of the following document types:

- "One pagers" (summaries of fact or perspective, limited to one page)
- Memos (more developed summaries, varying length)
- White papers (extensive reportage or analysis including evidence, in contrast to briefer memos or one pagers)
- Legislative concept proposals (outlines of model or idea or strategy for policy, without details)
- Legislative histories (reports of government action or inaction, based on government records)
- Committee reports (synthesis of committee decision and history of action on a topic)
- Speeches (to be delivered by elected or appointed officials)
- Testimonies (to be delivered by executives or professionals)

For some inside professionals, communication is the entire job. The communications director in the state budgeting scenario (chapter 1) is an example. A communications director is a generalist who:

- Writes and produces internal documents of many kinds
- Writes external public announcements of many kinds
- Produces kits of information for news media use

Other professional communicators in government are specialists. They include:

- Speechwriters who draft talks for officials to deliver
- Legislation writers who draft bills for deliberation and formulate laws for codification
- Debate reporters who produce stenographic transcripts and the published records of deliberation and debate

Professionals Outside Government.   Significant amounts of information used in policy making come from outside government. Experts of many kinds in academia, industry, and business write or contribute to white papers, reports of many kinds, and testimonies. In addition, professionals and managers in publicly regulated industries and businesses might provide needed information.

For some outside professionals, communication for public policy purposes is the main focus of their job. Lobbyists are an example. They are experts in a subject and are employed by organizations to ensure that policy makers have information about the subject that is germane to the interests of the employing organizations and to ensure that policy makers are exposed to the full range of arguments on a given issue. Lobbyists might brief legislators and their staffs, or they might draft legislation for consideration. Policy analysts are a different example. They may be either inside or outside government. They are experts in using quantitative and qualitative methods to examine problems and options for solving problems. Analysts might advise policy makers on the choice of policy instruments or provide research results to aid the formulation of policy.

Active Citizens.   Ordinary people in daily life inform and influence public policy making when they:

* Write or e-mail elected officials
* Provide comment on their experience relevant to a problem or a policy
* Testify about effects of a problem or a policy on their life or their livelihood
* Conduct letter-writing campaigns, create e-mail lists, and use phone trees
* Form a coalition of competing groups to cooperate in solving a problem
* Create a mechanism such as a lawsuit or a boycott to force response by institutional authorities
* Lobby as a representative of civic organizations, trade associations, professional associations, communities of interest, or constituencies

To summarize, in this chapter so far, you have been introduced to expectations, standards, roles, and communications associated with roles typically found in policy-making contexts. Change reading gears here, please. What follows next is a method, an outline of procedure. It is informed by the culture of public policy communication just presented. It consists of questions that translate culture into a routine intended to guide practical writing and speaking in a policy process. At the end of the outline are two checklists that translate standards into a compilation of expected qualities in documents or talks, intended to be used for assessing a document you have written or a talk you have planned.

Now, you should only read the outline and checklists for perspective and for familiarity. Later, when you are communicating (for instance, in a course or in meeting a real-life demand), use the outline (or method, as it is subsequently called) *before* you write. Use the checklists *after* you write.

A General Method of Communicating in a Public Process

Ask and answer the following questions to consider the context and situation for a communication, to plan it, and to produce it. The questions account for typical conditions of policy making as well as the dynamics of information exchange in a process. They prompt you to consider all the usual components and to take note of significant particulars that might affect your communication. Your answers to the questions should guide your writing and revising. (If you do not understand what some of the questions mean, it might be helpful to consult a standard textbook on professional or organizational communication.)

Practice this procedure literally (even if laboriously at first) until it becomes routine to ask these questions each time you have a need to communicate. At first, jotting down your answers and keeping your notes nearby as you write will be helpful. Later, when you habitually use this method to prepare for communicating, you can adapt it according to a specific demand. A word of caution: Even if you skip some questions, do not omit whole steps. All the steps are needed to cover the basics. Omitting a step in the preparation wastes time or causes other trouble later.

If you are writing for someone else or if you are producing a document with many contributors (the state budgeting example illustrates both), remember to consult with others as needed to answer the questions.

STEP 1: Prepare. First, ask questions about the policy process.

Policy
- To what policy process (underway or anticipated) does this communication relate?
- Does a policy already exist?

Problem
- What is the problem? What conditions are problematic?
- How do I define the problem?
- How do others define the problem?

Players
- Who are the players?
- What are their stakes in the process?
- Who else has a significant role in the process?

Politics
- What are the major disagreements or conflicts?
- What are the major agreements or common interests?
- How might players try to influence the outcome?

STEP 2: Plan. Second, ask questions about the communication.

Purpose
- Why is this communication needed?
- What do I want to accomplish?

Message
- What is my message?
- How does my message differ from others on the same topic?

Role
- What is my role in this process?

Authority
- Whose name will be on the document(s): Mine? Another's? An organization's?
- For whom does the communication speak?

Reception
- Who is (are) the named recipient(s)?
- Who will use the information?
- Will the document(s) be forwarded? Circulated? To whom? Represented? By whom?

Response
- What will recipients know after reading the document(s)? What will users of its information do?
- What is likely to happen as a consequence of this communication?

Setting and Situation
- What is the occasion? What is the time frame for communicating?

- Where, when, and how will this communication be presented?
- Where, when, and how will it be received? Used?

Form and Medium
- Is there a prescribed form, or do I choose?
- What is the appropriate medium for presentation and delivery? A written document? A telephone call? E-mail?

Contents
- What information will support the message?

Organization
- Where will a succinct statement of the message be placed?
- How should the contents be arranged to support the message?
- How will the document's design make information easy to find?

Tone and Appearance
- How do I want this communication to sound? What attitude do I want to convey?
- How do I want the document(s) to look? Is a style or layout prescribed, or do I choose how to present the contents?

Document Management
- Who will draft the document? Will there be collaborators?
- Who will review the draft? Who will revise it?

STEP 3: Produce. Based on your preparation and planning, write the document. Do it in three separate passes: draft first, review second, and revise third. Separating those tasks allows you to manage your time, handle distractions, and communicate better.

Draft
- Produce a complete working draft in accordance with your preparation and plan (answers to the questions above).

Review
- Compare the draft to the plan, and highlight any differences.
- Get additional review of the draft by others, if advisable.
- Refer to the checklists (shown below) to assess the draft's effectiveness and quality and to highlight needs for revision.

Revise
- Make the changes called for by review.

After a document or talk is produced, assess it by using the following checklists.

Two Checklists

Features of Effectiveness.   A public policy communication is most likely to be useful if it addresses a specific audience about a specific problem; has a purpose related to a specific policy process; represents authority accurately; uses the appropriate form; and is presented in a usable design.

- ☐ *Addresses a specific audience about a specific problem.*   In policy work, time is scarce. Specifying a communication's audience or intended recipient(s) and the subject or problem(s) saves thinking time for writer and reader (or speaker and listener). The information's relevance for the recipient should be made clear.

- ☐ *Has a purpose related to a specific policy process.*   Policy processes have several phases. Multiple policy processes are underway simultaneously. Timing matters. Agendas change. Stuff happens. Therefore, specifying a communication's purpose and relevance makes it more likely to get timely attention.

- ☐ *Represents authority accurately.*   Policy communications do more than present information; they also represent a type of participation and power. In order for a policy communication to be taken seriously, to have influence, and to influence rightly, the communicator's role and status—a citizen with an opinion, an expert with an opinion, a spokesperson for a non-governmental organization, a government official—must be accurately represented.

- ☐ *Uses appropriate form.*   Settings of policy work have their own conventions for communicating. Use the document type, style, and tone of presentation that are expected for the purpose and that accommodate working conditions in the setting of its reception.

- ☐ *Is designed for use.*   People's attention is easily distracted in settings of policy work. Dense, disorganized text will not be read or heard. For people to comprehend under conditions of time pressure and information overload, contents must be easy to find and to use. Written documents should chunk information, use subheadings, and organize details in bulleted lists or paragraphs or graphics. Spoken texts should cue listeners' attention with similar devices.

Measures of Excellence. No two communications are exactly alike, but every public policy communication should try to meet criteria for clarity, correctness, conciseness, and credibility.

- ☐ *Is clear.* The communication has a single message that intended recipients can find quickly, understand easily, recognize as relevant, and use.
- ☐ *Is correct.* The communication's information is accurate.
- ☐ *Is concise.* The communication presents only necessary information in the fewest words possible, with aids for comprehension.
- ☐ *Is credible.* A communication's information can be trusted, traced, and used with confidence.

Reference

Coplin, William D., and Michael K. O'Leary. *Public Policy Skills*. 3rd ed. Washington, D.C.: Policy Studies Associates, 1998.

❖ *3* ❖

DEFINITION:
FRAME THE PROBLEM

Scenario I

An undergraduate intern in a nonprofit, nonpartisan policy think tank concerned with Social Security, Medicare, and other social insurance programs is assigned to the director's office. The institute does not lobby or take positions on social issues. It produces influential analyses and reports for governments, the nongovernment public sector, and private foundations. The intern has done mainly routine office work before assignment to the director's office. In her new assignment, she uses spare time to read the institute's publications, browse its website, and become acquainted with its documents archive. One day she is asked to find the necessary information and to put together, by the end of the day, a list of the most pressing current issues for elder health care. She is helping the public policy director meet a deadline for producing a white paper on health care needs for an aging population. Even though the institute does not lobby, its publications are widely read by people in government. The white paper might influence a new administration's priorities for health care spending.

Scenario 2

A graduate student is chemically burned when an air bag deploys in an automobile accident. Emergency personnel and physicians are unfamiliar with the chemicals used in air bag manufacture, and they are unsure how to handle the victim's burns. Recovery leaves the student with lingering physical effects and with questions about air bag safety. To answer a list of questions generated by his own experience, he independently searches for consumer information on the materials in air bags. In scholarly publications by technical specialists and in science journalism, he finds warnings about potential injury from the force of inflating air bags, but no warnings about the potential for injury from their component materials. He makes notes on his initial findings.

Scenario 3

An undergraduate majoring in public policy studies assumes the role of a policy analyst in a classroom exercise. She chooses an environmental problem of concern to her, air quality. She studies the problem and uses statistical methods of cost/benefit analysis to consider varied policy options. As directed by the assignment, she reports her analysis in a memo for a legislator intended to help him or her develop an opinion.

In the first and third scenarios, students aid policy choice regarding a recognized problem. In the second scenario, a student follows up on a personal experience to identify an unrecognized problem. This chapter shows you how to define a problem and specify its issues for policy consideration and action.

How does public policy making begin? Typically, it starts with perception of a problem. Somebody perceives a condition in society or the environment to be wrong. Perceptions of a problem differ, so finding a solution often involves conflict.

Problems come to public attention in various ways. Sometimes the problem chooses you. Something happens, you are affected as a private individual, and you seek public action to address the problem. The triggering event might be large-scale, as in the attacks against New York City and Washington, D.C., in September 2001. After that event, families of victims organized to seek compensation or other action by government. In contrast, a triggering event might be small-scale, even personal and singular. Following her child's death due to a drunk driver, a parent formed the national nonprofit organization Mothers Against Drunk Driving to influence national law enforcement standards for drunk-driving. In another example, a scenario at the beginning of this chapter tells how a personal injury from air bag deployment motivated a student to learn more about auto safety regulations.

Sometimes you choose the problem. Choosers vary. For example, elected officials have authority to decide what is and is not a problem and which problems will receive attention. In the budgeting scenario (chapter 1), a governor and state legislative committees exemplify this kind of chooser. Authorities' choice processes might be aided by policy analysts acting as neutral experts to review policy options, as shown in the budgeting scenario. Interns in policy institutes or think tanks sometimes get involved in analysis, as shown in a scenario at the beginning of this chapter in which a student intern does research for analysis of needs for elder health care. A common type of chooser is the advocacy group that gets a problem into public debate or onto a legislative agenda.

To influence policy making, the perception of a wrong is not enough. If public policy is to be a solution, the wrong must be defined as one that policy makers can address. For example, you might perceive that obesity is wrong because it harms individuals, but individual obesity cannot be legislated. However, if you define obesity as a public health problem, you can relate obesity to public concerns such as health standards or medical research in the causes of disease. Those are problems with broad societal significance that can be addressed by policy makers.

Problem definition is important. As the logical first move of a policy process, problem definition sets the debate; it also predicts the solution. Different definitions lead to different solutions. For example, even though health authorities recognize obesity as a serious problem, the average citizen seemingly does not. Why, you wonder, are people in the United States getting fatter? Your question defines the problem differently, thereby revealing different potential solutions. By shifting the focus to people in everyday life, you expose another set of conditions relevant to obesity—behavioral issues such as eating habits, physiological issues such as genetics, cultural issues such as food preferences, and economic issues such as food industry profits. You point to solutions involving consumers, educators, businesses, and industries rather than doctors.

No matter how messy the process becomes, your action in a policy process is directed by your definition of the problem.

How to Define a Policy Problem

The goal, objective, scope, and product(s) of problem definition are specified below, along with a suggested strategy. Following, you will find two sets of guidelines for defining problems according to two different purposes: getting a problem on the public agenda (Purpose A) and aiding policy choice (Purpose B). Written products are suggested for each purpose: problem description (Purpose A) and policy analysis (Purpose B). Accordingly, apply the guidelines that fit your purpose and intended product.

> *Goal:* Ability to recognize problematic conditions and to define the policy problem they present
> *Objective:* Problem definition
> *Scope:* Individual or collective; local or broader in impact; well known or unrecognized; widely discussed or little considered; past, present, or anticipated
> *Product:*
>> For Purpose A, getting a problem on the public agenda:
>> • Written problem description with (or without) explanation of causes and with (or without) proposed solution
>> For Purpose B, aiding policy choice:
>> • Written policy analysis with (or without) recommendation
> *Strategy:* Provision of information necessary to your purpose

In formulating your strategy, expect to be flexible in the process. Problem definition can be iterative. After completing a task, you might find that you then must revise earlier work. Or, after defining a problem, you might find that you want to, or must, redefine it if conditions change or you gain more knowledge.

Purpose A: Get a Problem on the Public Agenda. You want to bring public attention to a problem of concern to you. It might be known to others, but only recently familiar to you. Or you might be aware of a problem of which others are unaware. In any case, you must understand the problematic conditions well.

To develop your understanding, follow an approach of observation and inquiry. Do the tasks below in sequence. Results of one task will help you perform the next one. Note: This task outline assumes that you are a novice in problem definition.

Task #1. Describe the problem and identify the stakeholders.

The first step is to describe the problem and name the concerned parties, or stakeholders. This involves recognizing problematic conditions, characterizing the problem that those conditions create, and specifying individuals as well as collectives that have a stake in the problem or its solution. To increase your awareness of the problem and to recognize public interests in it, you can proceed in any of the following alternative ways:

- Work from observation of experiences, practices, effects:
  - Note likes/dislikes about your (or others') daily routine
  - List good/bad aspects of your current or past job(s) or a family member's or a friend's job(s)
  - Sit for an hour in the office of a service provider to observe people affected by the problem and to observe the practices of policy implementers
  - Visit locales affected by the conditions or the policy to observe impacts on physical environments
- Work from subjective constructions:
  - Listen to or read stories (actual or imagined) that refer to the problem
- Work from unfinished business:
  - Reexamine a neglected need
  - Revive a former interest
  - Return to an incomplete project
- Work from anticipation:
  - Imagine the consequences if particular things continue as they are
- Work from ignorance:
  - Choose a matter that concerns others (but is unfamiliar to you) that you want to know more about

- Work from knowledge:
  - Consider the concern technically, informed by your (or others') expertise
- Work from values:
  - Consider the concern ethically or legally, informed by your (or others') ideals or commitments

Task #2. Specify the issues.

When a problem has been identified, it is not yet a policy matter until its issues for policy are specified. Issues refer to stakeholders' concerns, political disagreements, and value conflicts. To recognize issues, you might:

- Think about impacts of the problem. Who or what is affected by it?
- Conceive the problem narrowly and then broadly. Is it individual and local or more widespread?
- Conceive it broadly and then narrowly. Is it widely distributed or concentrated?
- Think about attitudes. How do different stakeholders perceive the problem? What values (ideals, beliefs, assumptions) are expressed in their definitions?
- Think about authority. How do stakeholders want to address the problem? Do they see government action as a solution? Do they agree or disagree on government's role?

Task #3. Offer solutions (if you are proposing a solution).

Solutions typically rely on policy instruments (actions such as spending more or spending less and starting or ending programs) that government can use. If you already have a positive and feasible solution to suggest, do so. (Generally, problem descriptions with a proposed solution get more attention.) If you need to think about it, if you want to counter a proposed solution, or if you want to create fresh alternatives, stimulate your thinking with any of these approaches:

- Review the problematic conditions with a fresh eye, looking for unnoticed solutions
- Reconsider a tried-but-failed or a known-but-ignored solution to find new potential

- Look at the problem from a different perspective (a different stakeholder's, for example)
- Assign it to a different governmental level or jurisdiction if government already addresses the problem
- Consult with nonprofit groups and nongovernmental organizations that are concerned about the problem
- Consider doing nothing (keep things as they are)

Task #4. Write the document: problem description and definition.

Before you write, use the method in chapter 2 to determine the rhetorical framework (audience, purpose, context, situation) for your communication. Then write with that framework in mind.

Problem descriptions can be presented in varied document types. If the type is prescribed for you, use it in accordance with your rhetorical framework. If you are free to choose the document type, choose one that fits your audience, purpose, context, and situation. Here are two options:

- Letter or essay describing problematic conditions, possibly identifying causes of the conditions
- Letter or essay conveying informed opinion, possibly advocating an approach to the problem

You might want to look at samples of each option. At the end of this chapter, you will find an essay written by a student. To see more essays and letters, go to these respected sources:

- *Congressional Quarterly* publications (*CQ Weekly Report*, *CQ Researcher*, or *CQ Annual*, available in print in subscribing libraries or online at http://www.libraryip.cq .com/home/home.jsp)
- Opinion sections of national newspapers such as the *New York Times* (http://www.nytimes.com), *Los Angeles Times* (http://www.latimes.com), *Chicago Sun* (http:// www.suntimes.com/index), or *Washington Post* (http:// washpost.com/index.html)

Do not choose to write a letter or essay simply because those forms are familiar to you, the writer. Consider your readers' needs and expectations, too. Are busy readers likely to be impatient with the loose organization of an academic essay or an informal letter? Might your audience want a succinct summary upfront? Might they want subheadings to announce the contents of each subsection? If the answers are yes, use those devices. They are borrowed from the inverted pyramid in news writing, and they are commonly used in professional communications. They work for public policy communication, too, because they aid in quick comprehension. As you read samples later in this chapter, notice their use.

Problem descriptions in any form are expected to answer the following questions:

- What are the problematic conditions? What problem do they cause?
- What are the issues for policy?
- What is your concern? What is your intended reader's concern?
- Who else is concerned (on all sides)?
- What are the key disagreements and agreements among those concerned?
- What plausible and realistic solution can you offer? (optional)

You must cite the sources to which your problem description refers. Use the citation style prescribed, or choose either the American Psychological Association (APA) style (described at www.apastyle.org) or the Modern Language Association (MLA) style (described at www.mla.org). Both APA and MLA style guides tell you how to cite a range of source types including government documents. (For example, see *Citing Government Information Sources Using MLA Style* at http://www .library.unr.edu/depts/bgic/guides/government/cite.html.) For further help with citing government sources, consult Garner and Smith's *The Complete Guide to Citing Government Information Resources: A Manual for Writers and Librarians* (rev. ed.).

After you write, check your document's quality against the checklists in chapter 2.

Purpose B: Aid Policy Choice. A problem is recognized. Policy alternatives for addressing it are under consideration. You are asked or wish to present a definition of the problem and to review the policy alternatives. Your intended audience might be policy makers, an interested community, or the general public. Follow a strategy of formal analysis using quantitative or qualitative methods. (Note: The task outline assumes that you are a novice in policy analysis. You are prepared to perform—outside the tasks listed here—appropriate technical analysis needed to answer the questions.)

Task #1. Identify the problem and the stakeholders.

- What is the problem? What brings it to attention?
- Why does the problem occur? What conditions lead to it?
- Whose behavior is affected, or whose concerns are relevant? Who are the target beneficiaries of solutions to the problem? Who are the implementers of policy to solve it?
- What stake does each (affected groups, target beneficiaries, implementers of policy) have in the problem?
- How does each define the problem?
- What ideals and values (equity, liberty, efficiency, security, loyalty) or ideologies (vision of how the world is or how it should be) are expressed in each definition?
- What conflicts of values or ideologies are evident among stakeholders?
- How does politics influence the problem?

Task #2. Specify alternative solutions and relevant criteria for evaluating them.

- What are the goals/objectives of a public policy to solve this problem?
- What policy instruments might achieve the goals/ objectives?
- What are at least two (alternative) policies to meet the need?
- What are the relevant criteria for choosing the best one? How do stakeholders weigh the criteria? How appropriate are the weights? What are the trade-offs among criteria?
- What would be the outcome of each alternative according to criteria you consider relevant?

Task #3. Recommend an alternative and explain your reasoning (if you are making a recommendation).

- Which policy option or instrument do you recommend? Why is it best? Why are other alternatives worse?

- What is the basis for your recommendation? What type of analysis supports it?
- How will your choice affect stakeholders?
- On what conditions (political, economic, organizational) does successful implementation of your choice depend?
- What are the constraints (political, economic, organizational) on implementing your choice?

Task #4. Write the document: policy analysis with (or without) recommendation.

Before you write, use the method in chapter 2 to frame your communication rhetorically and to plan it.

Policy analysis is communicated in varied document types. If an assignment prescribes a particular type, use it in accordance with your purpose and intended audience. If you are free to choose, you might use a memo.

You might want to look at sample policy memos. At the end of this chapter, you will find memos written by students in academic courses and by professionals in a government agency. To see other samples, go to this respected source:

- U.S. Government Accountability Office (GAO) (http://www.gao.gov/main.html). To search for GAO reports, testimonies, or opinions on your topic, go to Government Printing Office (GPO) Access (http://www.gpo access.gov/gaoreports/index.html)

Policy analyses in any form are expected to:

- Characterize a problem according to its size, scope, incidence, effects, perceptions of it, and influences on it
- Identify policy choices available to address the problem
- Offer perspectives to assist choice making
- Specify the basis for selecting any proposed recommendation (the type of analysis performed), the effects for different groups, and the factors that will affect its implementation

You must cite the sources to which your policy analysis refers. Use the citation style prescribed, or choose either the APA style (described at www.apastyle.org) or the MLA style (described at www.mla.org). Both APA and MLA style guides tell you how to cite a range of source types including government documents. (For example, see *Citing Government Information Sources Using MLA Style* at http://www.library.unr.edu/depts/bgic/guides/government/cite.html.) For further help with citing government sources, consult Garner and Smith's *The Complete Guide to Citing Government Information Resources: A Manual for Writers and Librarians* (rev. ed.).

After you write, check your document's quality against the checklists in chapter 2.

Four Examples

Example 1 (Purpose A—Get a Problem on the Public Agenda). Here's a problem description written by a student in an academic public policy writing course. It is an essay of informed opinion that advocates a chosen policy instrument.

OBESITY AND THE ROLE OF FAST FOOD

It's a sensitive subject, a faux pas in social conversation, but it's prime time for Americans to start talking about their weight. As health costs rise and Baby Boomers near the threshold of Medicare en masse, national focus must turn to the scales hidden in bathroom closets. Smoking and drinking are commonly blamed in courts of law and opinion as contributors to health problems, but obesity, an equally dangerous and more widespread affliction, is commonly overlooked.

Americans seem to know that fast food is not nutritious, but just how aware are consumers of the significance of what they put on their tray? Studies conducted by the Surgeon General and National Research Council found that 'cardiovascular disease, cancer, stroke, diabetes and obesity are linked to excessive intake of calories, fat, cholesterol and sodium.' These findings haven't stopped Americans from consuming larger portions of unhealthy prepared and fast foods. Why?

This may be attributed to a lack of nutritional content or ingredient labels on fast food packaging; there is nothing disclosing that a Happy Meal (burger, fries, and drink) has half the calories and all the fat and sodium an adult needs daily. There ought to be. If the regular consumption of a mass-produced, standardized, commercial product can lead to serious health risks, the consumer must be clearly advised. The present failure to do so is cause for concern.

Large health interest groups such as the American Heart Association and the American Cancer Society have long worked to educate Americans about proper diet and the health consequences of eating fast food. But they have had only limited success. The restaurant industry's powerful advertising and lobbying efforts overshadow health interest groups and continue to whet national taste buds and shape policy.

The most controversial topics being debated are the aforementioned nutritional and ingredient labeling of fast food products and whether fast food chains ought to be allowed in school cafeterias. In 1985 when nutritional labeling first became required on pre-packaged food, the FDA [Food and Drug Administration] favored the National Restaurant Association lobbyists who claimed that costs were prohibitive rather than listen to health officials or consult the public.

Government should not yield to business interests in matters of national health. State and federal governments must mandate the disclosure of nutritional information and ingredients on fast food packaging so that the public becomes better informed about the health risks of what they put on their plate. With easy access to information, consumers can better moderate their fast food choices and pressure restaurants to provide more nutritious alternatives.

Work Cited

Clark, Charles S. "The Fast Food Shake-Up." *CQ Researcher*, November 8, 1991: 838–43.

Reference

Bettelheim, Adriel. "Obesity and Health." *CQ Researcher*, January 15, 1999: 27–44.

What This Example Shows.   The document answers basic questions well (Purpose A, Task #4). In summary mode, the author situates the essay within a policy process by naming the problem, identifying key players, stating their positions, characterizing the politics, and advocating the author's position (see method, chapter 2).

The document's intended audience, reception, and use are not evident (see method, chapter 2). (The course assignment did not require the author to specify them in the document itself.) Consequently, although the document met an academic assignment's aims well, its real-world use is limited.

The document does not meet the standard for credibility. Two sources are probably insufficient to persuade recipients. (You cannot be sure, given unspecified audience and circumstances for this document's reception.) These ambiguities in the document reduce the writer's credibility (see checklist, chapter 2).

The writing meets the standard for conciseness. It is succinctly worded. It uses sentence structure well to manage the reader's attention. Well-placed short sentences emphasize the message. In longer sentences, references to the message are placed at the end where readers are likely to notice them more. Added italics show these techniques in the extract, below:

> There is nothing disclosing that a Happy Meal (burger, fries, and drink) has half the calories and all the fat and sodium an adult needs daily. *There ought to be.* If the regular consumption of a mass-produced, standardized, commercial product can lead to serious health risks, *the consumer must be clearly advised.*

This author uses the essay genre. That might work well enough, providing that the author knows who the readers are, why they want the information, and how they might use it (see method, chapter 2). It might work as an opinion published in a newspaper's editorial section.

Example 2 (Purpose A—Get a Problem on the Public Agenda).

This is a problem description written by a student in a public policy writing course. It is a memo describing a problem, identifying its causes, and enumerating proposed solutions. It proposes an alternative solution.

EDUCATIONAL FUNDING
IN NEW YORK STATE

Overview of the Problem

Nearly three million children attend New York State public schools, however these children do not all receive the same quality of education, or have access to the same resources. Because of inequalities in the way public schools are funded in New York, children in the public schools do not all have the same chance to succeed.

Causes of the Problem

Property Taxes

Approximately 54 percent of school funds are raised from local revenues, including a large part from property taxes (New York), however, property values vary greatly among neighborhoods (Koch). Schools in neighborhoods with low property values receive less money from their communities, thus creating a gap between what school districts in affluent and poor neighborhoods are able to provide for students.

State Funding Formulas

The formula New York State uses to distribute funds among public schools in the state does not take into account the differing needs of each district. For instance, New York City schools have a larger percentage of disabled students and English Language Learners than any other district in the state, yet these schools do not receive additional funding (New York).

Effect of the Problem

Low Student Achievement

According to a report about the educational status of New York State's school districts, there are major differences in student achievement between poor school districts, especially New York City, and the more advantaged school districts, mostly suburban schools. In 2000, only 42 percent of New York City public school students, in comparison with 84 percent of students from more advantaged districts, met state standards in Language Arts at the elementary level. Also, only 26 percent of New York City students earned a Regents Diploma, as compared to 65 percent of students from advantaged school districts (New York).

Proposed Solutions (Nationwide)

School Vouchers
School vouchers, which provide money for parents to send their children to private schools, have been put into place in some states. Opponents to school vouchers, such as the New York State United Teachers organization, claim that vouchers do not help the poorest students and serve to further weaken public schools ("Statement by New York State United Teachers").

The "Robin Hood" Approach
This policy, which is very controversial in some states, such as Vermont, takes money from the richest districts and gives it to the poorest districts (Koch).

Quality, Not Money
Some experts believe that giving more money to public schools will not necessarily improve the schools. Eric Hanushek, a professor of economics and political science at the University of Rochester, holds such a view. "We're pouring huge amounts of money into education," Hanushek said, "and we're not getting anything out of it" (Walters).

Next Steps

The debate on how to fund New York State's public schools is not one that will be easily resolved. A struggling economy and diverse views on how to improve the schools stand as barriers to progress on this issue. New ideas constantly need to be examined. One solution that is worth pursuing is the idea of taxing everyone in New York State at the same rate, then pooling that money and dividing it equally among school districts. This would eliminate the funding inequalities that result from raising revenue from local property taxes. However, this plan would draw criticism from the more affluent communities, who may not feel an obligation to help fund poorer districts.

Works Cited

Koch, Kathy. "Reforming School Funding." *CQ Researcher*, December 10, 1999. Vol. 9, No. 46 (1043–1063).
New York State. Campaign for Fiscal Equality, Inc. *In Evidence: Policy Reports from the CFE Trial*, 2000.
"Statement by United Teachers on U.S. Supreme Court Ruling on School Vouchers." 27 June 2002. *New York State United Teachers*, <http://www.nysut.org/media/releases/20020627vouchers.html> Sept. 2003.
Walters, Jonathan. "School Funding." *CQ Researcher*, August 27, 1993. Vol. 3, No. 32 (747–767).

What This Example Shows. Basic questions are answered (Purpose A, Task #4). In connecting this problem description to a policy process, the author omits to specify the players, although the problem, issues, key disagreements, and alternative solutions are specified (see method, chapter 2).

Like Example 1, this description exists in no specific communication situation. (The assignment did not call for it to be situated.) Although this description filled an academic assignment well, its real-world use is limited. It might work as a summary for members of a nonprofit organization or professional association of educators.

The author of Example 2 chose the memo genre that is commonly used for professional communications. Depending on how much the author knows about the readers and their purposes for the information, it might be the right choice. Many readers prefer a memo to an essay because memos are designed to be quickly read and easily understood. This problem description satisfies the preference. It meets standards for clarity and conciseness, not only by means of sentence structure but also by content organization and use of subheadings.

Example 3 (Purpose B—Aid Policy Choice). This is a policy analysis without recommendation written in memo form by a student in a public policy analysis course. (Its writer appears in the third scenario at this chapter's beginning.) The assignment that this memo fulfills specified the audience, purpose, and form for the communication. Students were asked to prepare a policy memo that a senator can use to develop his or her opinion on a problem chosen by the student.

333

POLICY MEMO ON AIR POLLUTION

Issues Surrounding Air Pollution Regulation

New Source Review (NSR) aims to reduce air pollution, specifically from large point sources. The 1977 Clean Air Act Amendments (CAA) created NSR, a permitting process that requires specified emission controls to sources on a case-by-case basis. To alleviate associated economic burdens, the CAA grandfathered older facilities, focusing its efforts mainly on new and modified sources. A modified source is subject to NSR if a physical or operational change yields "significant" increases in net emissions. In its attempts to reduce air pollution, the NSR program has been the subject of the following criticisms:

- The slow permitting process is inefficient in determining appropriate emission control technologies and when exemptions apply. Production is hindered by these delays and potential revenue may be lost.
- Mandated technologies are not a cost-effective use of the company's resources.
- There is no standardized definition of exemptions such as "routine maintenance."
- Maintenance exemptions extend the lifespan of polluting sources.
- Inconsistent calculation methods used blur indications of "significant" increases as administrations and agendas change.

With this in mind, President Bush developed a Clear Skies Initiative (CSI) that intends to reduce emissions using market-based solutions.

Because air is a shared resource with no specified ownership, air pollution exemplifies a classic "Tragedy of the Commons" scenario. Air pollution produces a negative externality for citizens across state lines adversely affected by reduced air quality. These differing regional impacts necessitate intervention on the national level. Industry pressures, with great impact on state and local government, extend less influence in the federal government.

Stakeholders

Numerous primary stakeholders have an interest in the issue. First, the Environmental Protection Agency (EPA) and state and local governments contribute resources to enforce NSR and stand to lose control of regulating industries' environmental practices. The EPA is responsible for administering the Clean Air Act, while states are responsible for meeting National Ambient Air Quality Standards (NAAQS). Lax air pollution standards would degrade air quality in the Northeast but would lower costs for Western and Midwestern industry. Large and small businesses and their trade associations are most interested in the economic burdens presented by implementing environmental technologies and the potential for reduced production capacity. Specifically utility companies, employees, and customers are concerned with the impact of compliance costs on jobs and utility prices. Furthermore, citizens and environmental organizations are interested in the impact of improved air quality on ecological and human health.

Policy Criteria

A policy to reduce air pollution at large point sources must fulfill the following criteria:
- Reduce air pollution to an optimal level. Calculating the total output of NOx, VOCs, and SO_2 emissions as a percentage of a determined baseline allows evaluation.
- Minimize compliance costs to ensure air pollution reductions are economically feasible and consumer prices are minimized. Assessment necessitates total industry cost estimates.
- Minimize permitting turnaround time to enable equal opportunities for compliance. This requires monitoring and estimating permit cycle times from the beginning of the application process to modification implementation.
- Equalize compliance opportunities for small and large businesses in an industry. Calculating compliance costs as a percentage of net income facilitates comparison among all industry firms.
- Reduce air pollution for citizens of the Northeastern U.S.

- Minimize compliance cost for the utilities in the Western and Midwestern U.S.

Alternatives to Current NSR Policy

1. Maintain status quo. The current system of government intervention is a form of "command and control" regulation. The EPA dictates what modifications constitute the need for NSR and directs them to the appropriate abatement technology. This policy alternative involves significant government regulation and oversight because the permitting process is determined on a case-by-case basis.

2. Maintain NSR with modifications. Redefine exemption rules. Under one alternative definition, NSR would apply when modification costs exceed set limits. Another definition would apply NSR when the portion of modification costs over total costs exceeds a set percentage. Additionally, create smaller, industry-specific oversight boards to speed up the review process and determine appropriate emission controls.

3. Cap and Trade program involves government set emission limits that decline over time. Facilities that reduce emissions to levels below specified caps could sell their extra emissions allowances at market price or could bank them for later use. Selling emission permits may motivate facilities to invest in innovative control technologies.

4. Create firm level emission standards that decline every five years. Each firm can decide which technologies are best to reduce [emissions] below the cap.

5. Government would levy an emissions tax. The administrative agency would calculate a money value for each ton emitted. Sources will be charged for each unit emitted to discourage air pollution. This alternative allows firms to decide when and how to reduce emissions.

What This Example Shows. The memo answers basic questions (Purpose B, Tasks #1, #2, #3). It fulfilled the aims of the academic assignment. It might also be useful beyond academia because it is situated (by assignment) in a specific context and communication situation (method, chapter 2).

The document generally meets standards for clarity and conciseness by means of content organization and sentence structure. The heavy use of abbreviations strains clarity, however. Unexplained abbreviations (for example, NOx) violate the standard of clarity altogether. It meets the standard for credibility by means of completeness (covers all necessary topics), word choice (uses special terminology understandably), and logical organization (arranges information persuasively for the intended reader). Due to the writer's attention to credibility, readers will likely presume the accuracy of the analysis (see checklists, chapter 2).

Example 4 (Purpose B—Aid Policy Choice).   This is a description of problematic conditions with recommendations. It is a report written by professional policy and program analysts in the U.S. Government Accountability Office (GAO), an investigative research and auditing staff of Congress.

MILITARY PAY: ARMY NATIONAL GUARD PERSONNEL MOBILIZED TO ACTIVE DUTY EXPERIENCED SIGNIFICANT PAY PROBLEMS

Highlights of GAO-04-89 (issued November 17, 2003), a report to the Chairman, Subcommittee on National Security, Emerging Threats, and International Relations, Committee on Government Reform, House of Representatives

Why GAO Did This Study

In light of the recent mobilizations associated with the war on terrorism and homeland security, GAO was asked to determine if controls used to pay mobilized Army Guard personnel provided assurance that such pays were accurate and timely. GAO's audit used a case study approach to focus on controls over three key areas: processes, people (human capital), and systems.

What GAO Found

The existing processes and controls used to provide pay and allowances to mobilized Army Guard personnel are so cumbersome and complex that neither DOD nor, more importantly, the mobilized Army Guard soldiers could be reasonably assured of timely and accurate payroll payments. Weaknesses in these processes and controls resulted in over- and underpayments and late active duty payments and, in some cases, largely erroneous debt assessments to mobilized Army Guard personnel. The end result of these pay problems is to severely constrain DOD's ability to provide active duty pay to these personnel, many of whom were risking their lives in combat in Iraq and Afghanistan. In addition, these pay problems have had a profound financial impact on individual soldiers and their families. For example, many soldiers and their families were required to spend considerable time, sometimes while the soldiers were deployed in remote combat environments overseas, seeking corrections to active duty pays and allowances.

The pay process, involving potentially hundreds of DOD, Army, and Army Guard organizations and thousands of personnel, was not well understood or consistently applied with respect to determining (1) the actions required to make timely, accurate pays to mobilized soldiers, and (2) the organization responsible for taking the required actions. With respect to human capital, we found weaknesses including (1) insufficient resources allocated to pay processing, (2) inadequate training related to existing policies and procedures, and (3) poor customer service. Several systems issues were also a significant factor impeding accurate and timely payroll payments to mobilized Army Guard soldiers, including (1) non-integrated systems, (2) limitations in system processing capabilities, and (3) ineffective system edits.

What GAO Recommends

GAO makes a series of recommendations directed at immediate actions needed to address weaknesses in the processes, human capital, and systems currently relied on to provide active duty pay and allowances to mobilized Army Guard soldiers. In addition, GAO recommends action, as part of the Department of Defense's

(DOD) longer-term system improvement initiatives, to ensure that reengineering efforts include the process, human capital, and systems issues identified in this report.

DOD concurred with GAO's recommendations and described actions recently completed, underway, and planned to correct the noted deficiencies.

This is a summary of the report. To view the full product, including the scope and methodology, go to http://www.gao.gov/cgi-bin/getrpt?GAO-04-89] GAO-04-89.

What This Example Shows.   Example 4 exhibits many desirable qualities in public policy communication. Although its authors did not use the guidance offered here to compose it, the document exemplifies well the qualities that this guide supports.

The report meets expected standards. It is intended for a specific audience whose particular information needs are identified in the document. It is clear, concise, and credible (see checklists, chapter 2).

GAO is a respected governmental agency, so the document gains some credibility from its source. In addition, the writers have used design features that support the assessment that the document's information is credible (and thus presumed to be accurate). By stating explicitly for whom the report speaks (GAO), to whom it is addressed (a congressional committee chair who requested it), and who has reviewed it (DOD, which is the subject of the investigation), the report is accountable. Its authority is traceable.

Attention is paid to likely circumstances of the document's reception. The organization highlights key information and makes it easy to find on referral (important if the document will be circulated or summarized in a briefing). For example, a one-sentence statement of the message is prominently placed under the pointed subheading "What GAO Found." In the document as a whole, subheadings highlight key information by asking implicit questions that are answered by the information immediately following. The subheadings function as signposts pointing to the location of important information. Within subsections, the presentation is top-down or general-to-particular order, with summaries first and details second. These organizational devices aid readers' rapid comprehension and meet information users' needs to recall information or to find it quickly in a document (see method, chapter 2).

The chosen genre is a report. Government agencies such as GAO often specify document types to be used for a purpose. In GAO's case, written reports are specified for presentation of formal (finished) results of any investigation performed at the request of a Congress member. Institutions sometimes also prescribe a house style. This report illustrates GAO style for report summaries, or the condensed abstract of a report: three main sections (why the study was done, what was found, what is recommended). The summary ends with a locator for the full report and contact information for the report's author. Both of these concluding devices add to the document's credibility.

Reference

Garner, Diane L., and Diane H. Smith. *The Complete Guide to Citing Government Information Resources: A Manual for Writers and Librarians*. Rev. ed. Bethesda, MD: Congressional Information Service, 1993.

❖ 4 ❖

LEGISLATIVE HISTORY: KNOW THE RECORD

Scenario I

When her supervisor in a policy think tank asks a student intern to specify unmet needs in elder health care, the supervisor gives no instructions on how to gather the necessary information. The intern considers how to approach the task. She knows that in order to identify unmet needs, she must know what current law provides. As a strategy for getting started, she works from familiar experience. She selects an issue that affects her elderly grandparents, nursing home care. She then starts to conduct research.

She goes, first, to the legislative reports produced by the think tank and published on its website. She finds that they are in-depth analyses of individual laws. Because the website has no index, she cannot locate all laws that refer to nursing homes unless she reads all the reports. She does not have time for that.

She then tries searching on the Internet, using an all-purpose search engine and the search term "nursing home care." She finds advertisements for providers and websites of advocacy groups, but little legislation or debate. She next asks the institute's professional staff for help. A policy analyst directs her to government databases and to commercial databases of government information on the Internet. She uses each database's indexing vocabulary to streamline her searches by subjects — first, nursing home care; then hospital care, prescription drugs, and other subjects as they emerge. Search results suggest to her that a good time frame to focus on would be the years in office of the previous federal government administration. She searches her favorite government database again in that time frame for legislation of particular interest, and she spends several hours reading summaries of laws and proposed bills. She also checks the final action taken on each.

By the end of the day, she writes a brief legislative history of elder care. She defines the most pressing current needs to be those that were recognized by the previous administration but that were left unresolved. She summarizes a list of needs culled from a range of bills or amendments proposed but not passed or adopted. She identifies the most significant failures (according to criteria that she provides) and distills the public debate surrounding them.

Scenario 2

The graduate student injured by air bag deployment searches federal legislative databases to find laws regarding automatic occupant restraints such as air bags and seat belts. He also searches records of deliberation and debate regarding air bags. He continues the research whenever he has the time, and he gets help from a librarian when he is stuck. He is careful (at the librarian's suggestion) to cite fully all sources as they are indexed in each database so that he can find the records again easily.

When he has no more time for research, he pauses to reflect. Research results confirm his hunch that consumer education got little attention and emergency services personnel training got no attention in the original legislation and surrounding debate on air bags. He writes a memo to himself that cites the original law, amendments, current regulations, and related records of debate.

Later, when he has time, he addresses a letter to the administrator of a federal agency for highway safety requesting amendment of the standard for air bag safety. To the letter he appends his legislative history of air bags.

> Public policy making requires information about prior government
> action. This chapter shows you how to research legislative records
> and to write a legislative history.

Many kinds of information are needed for policy making. To frame a
problem, identify its issues, or propose solutions, you might need to
know about influential social history, technological developments,
and economic patterns. You might consult scientific research, public
testimony, advice of expert consultants and lobbyists, statistical data,
government agency reports, transcripts of legal proceedings, and
more. But one kind of information is essential: the history of government action on the problem. To get that information, you must consult the legislative record; you must be able to conduct legislative research using government documents.

Why is knowledge of the record important? For policy process,
precedent is important. Action builds on prior action. Knowledge of
precedent helps you to frame problems and to find solutions. Context
is also important. The record shows deliberation and debate that tell
you about a law's original intent and the intent of amendments. If
you are proposing new action, credibility demands that you know the
history of prior action.

Who conducts legislative research, and for what purposes? Government staff (and sometimes interns) consult the legislative record
to help them frame problems and identify issues. Outside government, professional staff (and sometimes interns) in organizations of
many kinds, such as nonprofit groups, trade associations, and policy
institutes, consult the record. They do so in order to inform their advocacy or analysis. Similarly, active citizens consult the record as independent researchers. They might pursue a personal interest, or
they might volunteer to investigate a record of action that is relevant
to an organization's mission. For legal interpretation, court clerks, law
librarians, and legal services professionals regularly consult the legislative record to know a law's intent as part of adjudicating disputes
over a law's meaning.

Who writes legislative history documents? Often, the people who
conduct the research also write the document that reports the results.
Government staff or professional researchers on contract to a committee or agency or volunteers for organizations, as well as individuals doing independent research, might conduct research and report it
in a legislative history tailored to a particular need to know.

How to Conduct Legislative Research and Write a Legislative History

Goal: Knowledge of U.S. law regarding a defined problem based on consulting legislative records.

Objective: Credible reporting of government action.

Product: Written document tracing either history of a single law or history of laws on an issue.

Scope: Either a single law or an issue involving multiple laws. Relevant action might be at the federal, state, or municipal level, or at several levels. In addition to legislative records, administrative records of rule-making and regulation as well as judicial records of litigation might be needed.

Strategy: Multiple approaches available. No single approach to government records research fits all; however, you will save time and frustration by planning before you start. Use the following guidelines to select a strategy:

Know Why the Research is Needed. Legislative uses for the research might be satisfied with past records. In contrast, legal uses might require very current information not yet recorded that only an informant can provide. Knowing the purpose for the research tells you what, and how much, to look for. Will the information be used to make new law (legislative) or to interpret existing law (legal)?

In either case, there might be a published history that meets the purpose. Or you might need to find the records required to write a specialized history. Knowing the purpose for the research can help you (or a librarian assisting you) to decide where to look first. Do you want to find a history or write one?

Know the User and the User's Purpose for the Information. Who, exactly, will use the information, and what is his or her interest or need? The user might be you, gathering information for personal use or for an academic or internship assignment. Or the user for whom you are conducting the research might be a legislator who wants to amend an existing law. Knowing the user's purpose tells you what, and how much, to look for. Federal records only? State or municipal records also?

Set the Scope. Will the research follow a single law through all its forms and related actions—bill, codified statute, administration, regulation, amendment, and (possibly) adjudication? Or will your research follow an issue through policy changes and across multiple laws over time? What is the relevant time frame? What is the relevant level of government?

Take the Necessary Time, and Manage Your Time. Records research can take hours, days, or weeks, depending on how much you already know, what you are looking for, where the records are, how well you have planned, and other contingencies. Prepare for the reality that legislative records research will take time, probably more time than you initially planned. What is your deadline for completing the research? What is your schedule for conducting the research and writing the necessary documents?

Use Existing Skills, and Add Needed Ones. If you have a well-defined problem, are willing to learn, are curious and persistent, and have basic research skills including ability to ask questions, identify relevant sources, and search computer databases, you are basically ready to perform legislative research.

You might need to learn about the legislative process, government record types, and standard tools for researching government records. If so, review as necessary, using the tools suggested below.

Task #1. Review the legislative process.

If you already know federal legislative procedure well or if you are tracing state law, omit Task #1 and go on to Task #2.

As you conduct research in government records, you can feel as if you are drowning in information, classification systems, procedure names, and document types. Also, if you start into records searching without knowing the underlying legislative process, you will quickly become lost. Use the following reviews of the process to revive your effort (bookmarking your favorite and returning to it as often as needed):

- The House: How Our Laws Are Made (by House of Representatives Parliamentarian)
  http://thomas.loc.gov/home/lawsmade.toc.html
- The Senate: Enactment of a Law (by Senate Parliamentarian)
  http://thomas.loc.gov/home/enactment/enactlawtoc
  .html

- The Legislative Process (by House of Representatives Information Office)
  http://www.house.gov/house/Tying_it_all.html
- The Legislative Process (by Indiana University Center on Congress)
  http://congress.indiana.edu/learn_about/legislative.htm
- The Legislative Process (by Capitol Advantage)
  http://congress.org/congressorg/issues/basics/?style=legis

Task #2. Conduct research.

Do you want to find a history or write one? Decide early whether your purpose is served by using an already published history or by producing one. For single laws, commercial research services such as the Congressional Information Service publish legislative histories with varying levels of detail. To look for a published history for a single law, try these sources:
- Law Librarians Society of Washington, D.C., Legislative Sourcebook
  http://www.llsdc.org/sourcebook/fed-leg-hist.htm
- CIS/Annual (Year), Legislative Histories of U.S. Public Laws

You are unlikely to find published legislative histories for an issue. They are typically produced by, or for, the people who want the information.

As a rule of thumb, federal records are generally accessible online and in research libraries. State records are generally less so, but an individual state's records might be available online or, more likely, in the print archives of the state's library. Local government records are generally not available without going to the municipality to ask about access to records. Few municipalities put their records online.

Major tools for finding federal and state records are provided by government information services, either free or by subscription. Free services can be accessed by any computer with World Wide Web access. Subscription services are accessed via the Web by authorized users of facilities provided by a subscriber such as a university library.

From your computer at home and in many public libraries, you can freely access federal records back to 1970 (and link to online state records) through:

- Thomas (Library of Congress)
  http://thomas.loc.gov
- Government Printing Office (GPO) Access
  http://www.access.gpo.gov/su_docs/index.html

Other free portals to numerous federal government websites that link to records are:

- Federal Web Locator (Information Center at Chicago–Kent College of Law, Illinois Institute of Technology)
  http://www.infoctr.edu/fwl/index.htm#toc
- Federal Government Documents on the Web (University of Michigan Documents Center)
  http://www.lib.umich.edu/govdocs/federal.html
- U.S. Government Documents (Mansfield University)
  http://www.mnsfld.edu/depts/lib/gov.html

For state legislatures and local government, these are good portals:

- State and Local Governments (Library of Congress)
  http://lcweb.loc.gov/global/state/stategov.html
- Government Resources on the Internet: State, Local, International (Virtual Chase)
  http://www.virtualchase.com/govdoc/jurisdiction.html
- Law Librarians Society of Washington, D.C., Legislative Sourcebook: State Legislatures, State Laws, State Regulations
  http://www.llsdc.org/sourcebook/state-leg.htm

An excellent subscription service that librarians consult for comprehensive federal legislative information is Congressional, based on the (print) *Congressional Information Service* (*CIS*) at http://web.lexis-nexis.com/congcomp.htm.

Both free and subscription services are available in federal depository libraries. Those are libraries that make GPO materials publicly available in the library's region. Find a depository library near you in the Federal Depository Library Locator at http://www.gpoaccess.gov/libraries.html.

Libraries offer a special advantage: librarians! For professional, skilled, and time-saving assistance in legislative research, always ask a librarian.

General Tips for Using Government Information Libraries

- Depository libraries have federal government records in all available forms—digital, print, and microfiche. Depending on what you want to know, you might need all three. Online access to digital records is convenient for recent records, but print and microfiche are still important, too, for several reasons. Records prior to the 1970s are not yet available online, and some will never be. You can miss a lot of legislative history if you only search online. Also, print compilations are sometimes easier to use, because they are well supplemented by indexes and other locator aids.

• When using a tool new to you, check first for finding aids such as an index. You will save much time this way. (Note: Subscription services have more indexes than do free services.)

• You should take detailed notes as you go. Jot down contextual information as well as target information. List names of people, committees, subcommittees, and bill or law citations mentioned in the target record. Why? If your first search method fails, these notes can restart your search; they give you alternative ways to search.

• You can use what you know to find what you want. For example, if a student intern researching elder health care jots down key terms, citations, names, and dates as she works in a database of government records, she is prepared to search by any of these alternatives:

 – By subjects discussed in the record (for example, elder
   health care)
 – By citation (number and letter "addresses") of a particular legislative record in a system of citation, (for
   example, H.R. 1091–106 for a particular House of
   Representatives bill)
 – By names, dates, committees, or other elements of a legislative process (for example, the name of the senator
   sponsoring a bill)

In other words, she could find legislation on elder health care by subject (elder health care, nursing home care, Medicare, and so on), or by citation (H.R. 1091–106), or by legislative process information (Senator Ted Kennedy; hearing witness Donna Shalala, Department of Health and Human Services; Senate Committee on Health, Education, Labor, and Pensions).

Task # 3. Write the legislative history document.

To write your legislative history, begin by using the method in chapter 2. You can reuse the thinking that went into planning your research (see "Strategy" under "How to conduct Legislative Research and Write a Legislative History" earlier in this chapter). Use it to plan your legislative history document. Let your intended reader's needs for the information guide your selection of information for the history.

What is the message of a legislative history? It is your conclusion formed after consulting the record. The history's scope is set by the purpose (whether you are writing a law history or an issue history) and by the amount of information required to support your message. In any case, you must organize your information to support the message. Organizational options include chronology (to show developments over time), significance (to highlight influential legislation), and trend (to show a pattern).

If no form is prescribed for presenting the results of your research, you might choose to use the following standard reporting format for professional and technical communication:

- Overview that concisely summarizes both the message and the key information in the document
- Subsections that provide summaries of information
- Subheadings that label each subsection
- Citations that are provided for each subsection

Citations are very important in a legislative history. Both the history's credibility and the user's (as well as the researcher's) convenience demand that all sources be easy to relocate for confirmation and referral. Citations are the means of doing so. A full citation provides three kinds of information about a source: what type of record it is, how it is classified in a system of documentation, who publishes it (a commercial research service or government). For government records, a full citation includes all the elements that help to identify a source. In legislative research, a full citation, or government style, is preferred over a terse citation, or legal style, that provides only an abbreviated source identifier, number in a system of documentation, and date. If either the government style or legal style is prescribed for you, use that style. If not, choose the appropriate style and use it exclusively. Do not mix styles.

Here is a list of the elements in a full citation, or government style, for citing federal or state legislation:

- Issuing agency (house, number, session, year)
- Title (document number and name; long name may be abbreviated)
- Edition or version
- Imprint (city, publisher, date of publication)
- Series (serial list of publications)
- Notes (in parentheses, add anything not already included in the citation that helps to locate the document)

Following are two illustrations of government style:

1. U.S. House. 101st Congress, 1st Session (1989). H.R. 1946, A Bill to . . . Authorize the Department of Veterans Affairs (VA) to Provide Home, Respite, and Dental Care. Washington: Government Printing Office, 1990. (GPO Microfiche no. 393, coordinate C13.)
2. U.S. House. 104th Congress, 1st Session (1995). "H.R. 3, A Bill to Control Crime." Version: 1; Version Date: 2/9/93. (Full Text of Bills: Congressional Universe Online Service. Bethesda, MD: Congressional Information Service.)

In the second illustration, the final element shows that the source is proprietary, or a commercial research service publication available to paying subscribers.

If you need more help on citing, see:

- Diane L. Garner and Diane H. Smith, *The Complete Guide to Citing Government Information Resources: A Manual for Writers and Librarians* (rev. ed.) (Bethesda, MD: Congressional Information Service, 1993)
- Citing Government Information Sources Using MLA (Modern Language Association) Style at http://www.library.unr.edu/depts/bgic/guides/government/cite.html
- Uncle Sam: Brief Guide to Citing Government Publications (University of Memphis Depository Library) at http://www.lib.memphis.edu/gpo/citeweb.htm
- How Do I? Cite Publications Found in Databases for Thomas (Library of Congress) at http://thomas.loc.gov/tfaqs/16.htm

Remember to check your final product against the standard (see checklists, chapter 2).

Two Examples

Example 1.   Here is a student's legislative history written for a public policy writing course assignment. For the related problem description, see Example 1 at the end of chapter 3.

THE LEGISLATIVE HISTORY OF NUTRITIONAL LABELING

Overview

Government has debated the topic of food labeling for nearly 100 years. Its history of legislation passed and court cases settled shows where we've come from and sets a precedent for future legislation. In 1906 Congress was concerned with establishing a basic standard for product labels to prevent consumers from being misled. Since then changes in science and public opinion have necessitated drafting new bills that fill gaps in legislation and place more restrictions on product labels to better protect and inform consumers. In the late 1980's that meant requiring nutritional labels on pre-packaged grids listing calories, fat, sugars, and other food values to inform an increasingly health-conscious America. Now in 2002, America's growing taste for increased portions of unhealthy fast food must be addressed by filling the gap in the Nutritional Labeling and Education Act exempting fast food from nutritional labels.

Major Legislation and Legal Decisions

59th Congress

H.R. 384: "The Food and Drug Act of 1906." This act was the first on record in the United States that governed the contents of product labels. The legislature was concerned that manufacturers and distributors were labeling their products in a manner that misled consumers. Product labels that falsified ingredients or other product information were considered "adulterated" by the act. Through this legislation, food and drugs were required to be labeled with "distinctive names" that pertained directly to their contents and to have those names and the manufacturers' locations printed clearly. To enforce this bill, the Department of Agriculture was empowered to inspect, on demand, all packaged goods manufactured or transported within the United States, levying fines on violators.[1]

76th Congress

S.5: "The Federal Food, Drug and Cosmetic Act of 1938." This legislation was intended to replace the Food and Drug Act and cover a greater variety of products, including cosmetics, with more specific language that clarified vagueness in the 1906 act. The new act regulated items on store shelves (an important addition), broadened the definition of "adulterated" to include spoiled or mishandled food, and placed tighter restrictions on how food could be labeled. If products claimed to serve specific dietary needs or produce certain health benefits, their labels had to contain a list of ingredients and be approved by the Secretary of Agriculture. The Secretary could now freely inspect not only the goods themselves, but also any factory, warehouse, or establishment that produced, stored, or sold them and freeze the sale of products that could be considered "adulterated." This was deemed much more effective than fines in deterring violators.[2]

85th Congress

H.R. 13254: "Food Additives Amendment." Created to amend the Federal Food, Drug and Cosmetic Act to cover food additives. This amendment shows government recognition of a growing trend in the food industry to use food additives and flavor enhancers with possible adverse health effects in order to lower costs. This amendment requires that before any food additive or flavor enhancer is used, its producers must disclose the additive's chemical composition and the results of a certified health study attesting to the additive's safety in the specific dosage. The effect was a dramatic decrease in the use of sodium and its derivatives as preserving agents.[3]

95th Congress

S.1750: "Saccharin Study and Labeling Act of 1977." This legislation is an extension of the Food Additives Amendment that called for the study of a possible link between saccharin consumption and cancer. At time of passage, saccharin, a sugar substitute, was a tremendously popular product and the implication that its usage could cause cancer was serious. The study found a conclusive link between saccharin usage and increased incidence of cancer in laboratory animals but it could not convince legislators there was a significant risk to humans. Instead of upsetting the marketplace based on "inconclusive" results, the Health, Labor, Education, and Pensions Committee implemented mandatory labeling. All products containing saccharin must clearly state: "Use of this product may be hazardous to your health. This product contains saccharin which has been determined to cause cancer in laboratory animals." What makes this bill noteworthy is that legislators approved of allowing an ingredient with alleged health risks to remain on the market *provided that it had* a clearly stated health advisory on the packaging.[4]

96th Congress

S.1196: "Disease Prevention and Health Promotion Act of 1978." The applicability of this legislation is its position on the effectiveness of disease prevention programs. The Committee on Health, Labor, Education, and Pensions found that contrary to popular opinion, "Americans are not fully informed about how to improve their own health and want more knowledge, that Federal, State and local governments have a role to play in providing that information, and that government at all levels has the capacity and the responsibility to help communities and individuals reduce the burden of illness through the prevention of disease and the promotion of good health."[5]

99th Congress

S.541: "Nutrition Information Labeling Act of 1985." A bill to amend the Federal Food, Drug and Cosmetic Act to require that a food's product label state the specific, common-name and the amount of each fat or oil contained in the food, the amount of saturated, polyunsaturated, and monounsaturated fats contained in the food, the amount of cholesterol contained in the food, and the amount of sodium and potassium contained in the food. This is the first bill to require nutritional labels, although it does not cover restaurants or raw agricultural products.[6]

99th Congress

H.R. 6940: "Amend Food, Drug and Cosmetic Act." This bill requires baby formula to contain a prescribed nutritional content in order to be sold in the US. The significance is that government recognizes the need to not only disclose nutritional content but also regulate that content in order to ensure the well-being of the consumer.[7]

Arbitration

In response to two petitions filed by The Center for Science in the Public Interest, New York State filed suit against McDonalds Corp. alleging that their Chicken McNuggets were not the "pure chicken" advertised. In an out-of-court settlement McDonalds Corp. agreed to withdraw the ads and disclose the ingredients and nutritional content of their menu in pamphlets and posters at their New York restaurants. At the same time attorney generals in ten other states began the process of filing suit to require nutritional and ingredient disclosures from McDonalds and other major fast food chains. In a national settlement still in effect, McDonalds, Burger King, Jack in the Box, Kentucky Fried Chicken, and Wendy's agreed to offer separately printed nutritional information in pamphlets or on posters in stores around the country. This action resulted in McDonalds cutting back on beef-frying and discontinuing the use of yellow dye No. 5 that has been known to trigger allergies. However long-term compliance with the settlement has been inconsistent and only Jack in the Box has made information consistently available nationally. The rest of the chains only provided them in an average of 33 percent of locations.[8,9]

100th Congress

S.1325: "Fast Food Ingredient Information Act of 1987." This bill was written in response to a greater nutritional consciousness and the national settlement mentioned in the lawsuit above. The bill sought to amend the Food, Drug and Cosmetic Act to force fast food restaurants to label pre-packaged goods with nutritional labels and to display nutritional and ingredient information in clearly visible places in their restaurants. The bill also sought to amend the Federal Meat Inspection Act and the Poultry Products Inspection Act to allow for nutritional information to be posted in restaurants. President Reagan vetoed this bill because of possible, negative economic consequences.[10]

101st Congress

H.R. 3562: "The Nutrition Labeling and Education Act of 1989." An amendment to the Food, Drug and Cosmetic Act designed to expand on the requirements of the Nutritional Labeling and Education Act. The bill states that food will be deemed misbranded unless its label contains: serving size, number of servings, calories per serving and those derived from fat and saturated fat, and the amount of cholesterol, sodium, total carbohydrates, sugars, total protein, and dietary fiber per serving or other unit. Authorizes the Secretary of Health and Human Services to require additional label information.[11]

Sources

1. U.S. House of Representatives. 59th Congress. 2nd Session (1906). "H.R. 384 Food and Drug Act of 1906." Washington Government Printing Office, 1981.
2. U.S. Senate. 76th Congress. 1st Session (1938). "S.5 Federal Food, Drug and Cosmetic Act of 1938." Washington Government Printing Office, 1981.
3. U.S. House of Representatives. 85th Congress. 2nd Session (1958). "H.R. 13254 Food Additives Amendment." Washington Government Printing Office, 1981.
4. U.S. Senate. 95th Congress. 2nd Session (1977). "S.1750 Saccharin Study and Labeling Act of 1977." Washington Government Printing Office, 1978 (Thomas Bill Summary S.1750).
5. U.S. Senate. 96th Congress. 1st Session (1978). "S.1196 Disease Prevention and Health Promotion Act of 1978." Washington Government Printing Office, 1980 (Thomas Bill Summary S.1196).
6. U.S. Senate. 99th Congress. 1st Session (1985). "S.541 Nutrition Information Labeling Act of 1985." Washington Government Printing Office, 1985 (Thomas Bill Summary S.541).
7. U.S. House of Representatives. 99th Congress. 2nd Session (1986). "H.R. 6940 Amend Food, Drug and Cosmetic Act." Washington Government Printing Office, 1986 (Thomas Bill Summary H.R. 6940).
8. Clark, Charles S. "The Fast Food Shake-Up." *CQ Researcher,* November 8, 1991: 838–43.
9. *McDonalds to Introduce Nutrition Information Programs in New York: Press Release.* 30 April 1986. New York: New York State Attorney General's Office Consumer Protection Bureau.

10. U.S. Senate. 100th Congress. 1st Session (1987)."S.1325 Fast Food Ingredient Information Act of 1987." Washington Government Printing Office, 1987 (Thomas Bill Summary S.1325).
11. U.S. House of Representatives. 101st Congress. 1st Session (1989). "H.R. 3562 The Nutrition Labeling and Education Act of 1989." Washington Government Printing Office, 1989 (Thomas Bill Summary H.R. 3562).

What This Example Shows. This history of an issue includes litigation as well as legislation regarding the issue (see Scope, this chapter). It is designed as a report, with an initial overview followed by summaries of major legislation arranged chronologically. Subheadings (congressional session and date) move the chronology along. Each summary concludes with a statement of the act's significance in a pattern or trend. The message of the report is to show the trend (see method, chapter 2).

The title and overview connect this report to a context, the policy process of amending existing legislation. No audience is specified; the course assignment did not require the specification. That is a limitation, but the history could nonetheless meet expected standards. It could be a report for a nonprofit organization or an advocacy group, which might use it to inform citizens, group members, or legislators (see method, chapter 2).

The author is a convincingly informed advocate. Credibility is enhanced by the report's organization, which suggests knowledgeable authority. The author has recognized a legislative trend and has condensed and ordered relevant legislation as well as litigation to highlight milestones in that trend.

Careful citation also supports credibility. Readability is served well by the way citations are handled. Citations are found in two locations in the text. Subheadings for summaries cite the legislative session, record number, and common name of each bill; a footnote at the end of each summary refers to citations at the end of the document, where the bill is fully referenced using government style (see Task #4, and checklists, chapter 2).

Example 2. Here is another student's legislative history of an educational reform. This history was also written for a coursework assignment.

CHARTER SCHOOLS IN AMERICA

A charter school is a public policy designed to ameliorate many problems in American public education, such as overcrowded schools, poor student achievement, lack of motivation and innovation by teachers, oppressive and rigid policies and regulations, and lack of parental or community involvement in schools. While discussion of educational "charters" for teachers to explore innovative approaches to teaching began in the 1970s, there was no formal legislation until Minnesota created the first state charter school policy in 1991. In the last ten years, 36 states have also drafted individual legislation which projects varying degrees of uniformity between charter schools across the country. While federal legislation regarding charter schools is not broad, the federal government did begin making provisions for the creation and funding of charter schools in 1998 and has revised their policy as recently as January 2002.

Landmark State Legislation

Minnesota Statutes Chapter 124D.10 Charter Schools (1991, Updated Annually)
The Minnesota charter school legislation was ground breaking because it was the first state legislation passed regarding charter schools. The document specified that a charter must meet at least one of the following requirements:
- improve student learning;
- increase learning opportunities for students;
- encourage the use of different and innovative teaching methods;
- require the measurement of learning outcomes and create different and innovative forms of measuring outcomes;
- establish new forms of accountability for schools; or
- create new professional opportunities for teachers, including the opportunity to be responsible for a learning program at the school site.

More than 80% of states creating charter school legislation following Minnesota based the required purposes for charters in their state on the previous list. Minnesota continues to be distinct from other legislation in that it specifies that in an open-enrollment school (a start-up school as opposed to a conversion school) the student body must reflect the racial composition of the school's neighborhood. The legislation also liberally allows for a wide variety of individuals and organizations to apply for an unlimited number of charter schools.[1]

Arizona Revised Statutes Title 15 Chapter 1 Article 8 (1994).
Arizona's charter school law allows both the state and local boards of education to grant charters independently of one another providing for a greater number of charters to be accepted. This may be one reason Arizona (with 433 schools) has 58% more charter schools than the next greatest amount of schools in a state (California with 274 schools). The legislation also provides the longest term for charters to be granted (15 years). Arizona is one of the leading states in autonomy for both students and teachers. Arizona is among the most lenient of states with charter school laws, which makes it among the most progressive as well.[2]

Appendix 1: Charter School Legislation by State and Comparative Factors [Detail not shown here.]

Landmark Federal Legislation

105th Congress: H.R. 2616 "Charter Schools Expansion Act of 1998."
Introduced: 6 October 1997   *Sponsor:* Frank Riggs (R-CA)
Final Status: Signed as Public Law 105–278 (22 October 1998)

This is a bill to amend Titles VI and X of the Elementary and Secondary Education Act (1965) to improve and expand charter schools. There are several significant highlights of this legislation:

- Governing bodies of public charter schools must have a contract with their authorizing agency from which to base the continuance or revocation of their charter.
- State education agencies (SEAs) are authorized to use federal funds for the planning, designing, and initial implementation of public charter schools.
- Secretary of Education is given reserved funds to make sure charter schools receive the appropriate funding from their local education agencies (LEAs).

- SEAs and LEAs can receive grant funds based on the status and quality of charter schools in their state, their provisions for students with disabilities, their dissemination of best practices to other charter schools, and knowledge of federal funding formulas.[3]

107th Congress: H.R. 1 "No Child Left Behind Act of 2001"
Introduced: 3 March 2001   *Sponsor:* John A. Boehner (R-OH)
Final Status: Signed as Public Law 107–110 (8 January 2002)

Public Law 107–110 was signed into law as the most aggressive education reform the United States has seen in several decades. Title V, Subpart B, of this legislation authorizes $300 million to help states allocate funding for the initial implementation of charter schools, the evaluation and assessment of their achievements, and the costs of their physical facilities. The power of the Secretary of Education to provide grants to SEAs is increased so grants can be authorized for 3-year terms based on criteria such as: the ambitiousness of the charter; the likelihood that the charter will increase student achievement; the priority of the charter school for disadvantaged populations of students; the quality of evaluation and assessment plans set forth in the charter; and the flexibility of the SEAs and LEAs to accommodate charter school initiatives. This title also mandates that states must review charters at least every five years, if not more frequently.[4]

State by State Sources of Information on Charter Schools in thirty-seven States
[Detail not shown here.]

General Sources Consulted

Education Week on the Web. "Charter Schools." www.edweek.com/ context/topics/issuepage/charter.html (January 2002).
Hassel, Bryan C. *The Charter School Challenge: Avoiding the Pitfalls, Fulfilling the Promise.* Brookings Institution Press. Washington, D.C.: 1999.

Legislation Cited

1. Minnesota Statutes 2001 Chapter 124D.10 Charter Schools. Online. www.revisor.leg.state.mn.us/stats/124D/10.htm (24 January 2002).
2. Arizona Revised Statutes Title 15 Chapter 1 Article 8. Online. www.azleg .state.az.us/ars/15/title15.htm (5 February 2002).

3. "Charter Schools Expansion Act of 1998." P.L. 105–278, *United States Statutes at Large.* 112 Stat. 2683.
4. U.S. House. 107th Congress 1st Session (2002). H.R. 1 "No Child Left Behind Act of 2001." Washington: Government Printing Office, 2002 (Thomas Bill Summary. Online. P.L. 107–110).

What This Example Shows. This issue history traces state and federal legislation. It emphasizes state records because most of the action on the issue has been at the state level. The document's purpose is to inform; no action is called for. The history might serve as a background summary to inform civic groups, professional educational associations, or governmental audiences (see method, chapter 2).

The course assignment for which this example was written did not require the specification of a context or a particular policy process. An unspecified audience and purpose might undercut the real-world usefulness of the report. The author might be asked "So what?" or "Why do I need to read this?" To preempt such reactions, a title, a header with a subject line, or a revised overview might pointedly relate this history to the interests of particular information users. For instance, this history might be titled to catch the attention of a parent-teacher association that is advocating a change in a state's charter school law (see method, chapter 2).

The author is convincingly informed. However, because the summaries lack analysis or any statement of an act's significance, the writer's authority is undercut.

The document is concisely organized and well formatted. Sentences, however, are typically long and many sentences include unnecessary words (see checklists, chapter 2). The writer could shorten sentences to emphasize key information better, as illustrated below.

Original

Public Law 107–110 was signed into law as the most aggressive education reform the United States has seen in several decades. (21 words)

Revised

Public Law 107–110 was the most aggressive education reform in decades. (11 words)

The revision avoids redundancy ("Public Law" and "signed into law") and removes unnecessary explanation ("the United States has seen").

❖ 5 ❖

POSITION PAPER:
KNOW THE ARGUMENTS

Scenario I

A student government representative wants to change the culture at her university to discourage drug and alcohol abuse. As a dormitory resident advisor, she knows firsthand that campus culture encourages recreational drug use and underage as well as binge drinking. As a member of the Judicial Affairs subcommittee of the student assembly, she accomplished revisions in the university judicial system to increase sanctions against drug and alcohol use as well as penalties for violation of the sanctions. However, the sanctions and penalties have had little impact on the character of campus life. In a report that she authors for the student assembly's Judicial Affairs subcommittee addressed to the Dean of Student Affairs, she argues that judicial action is not enough. She cites evidence from dormitory life based on her resident advising experience. She claims that comprehensive action is needed to reduce dependence on drugs and alcohol for social interaction. She specifies needs to update university policy, to reorganize administration of campus life, and to design educational interventions. In her choice of proposed solutions, she has anticipated opposing arguments by other student government leaders and by some university administrators favoring either the status quo or increased sanctions and enforcement. Her purpose as principal author of the subcommittee report is to deepen the campus debate on drugs and alcohol by focusing on the central question of why campus life encourages their use.

Scenario 2

At the same university, another student majoring in public policy studies serves as an officer of a national student association that advocates better drug policy. In that role, she writes a memo to the director of a national drug control policy institute stating the association's position on recent legislation and asking the director to rethink the policy institute's support for the 1998 amendments to the Higher Education Act of 1965. Those amendments, titled the Domestic War on Drugs and Higher Education Act Amendments of 1998, barred students with drug-related convictions from receiving federal financial aid for education unless they undergo rehabilitation. The student leader claims that the amendments are a flawed approach to reducing drug use and that the implementation will not work because no funding is provided to pay for rehabilitation. She anticipates that the director of the drug control policy institute will disagree with her organization's position. She anticipates, too, that he will not address the absence of funding for rehabilitation. By focusing her argument on this legislation's wide impact and its flawed design for implementation, she intends to broaden the national debate on effective mechanisms for controlling drug use and to draw attention to government's responsibility.

Making public policy requires making arguments and understanding arguments. This chapter helps you to argue a position, to critically analyze other arguments, and to recognize grounds for cooperation as well as competition among arguments.

A policy argument supports a claim that something should or should not be done. Such arguments have two main components: a claim and its support. The claim asserts what should or should not be done.

Or it takes a position on a debated question. The support for the claim presents the facts, interpretations, and assumptions that lead to making that claim. The argument's presentation should be intentionally constructed to convince others to accept the claim and to agree with the position.

Arguments are made both implicitly and explicitly. For example, to define obesity as a health problem is to argue implicitly that obese people are potentially or actually unhealthy. To trace legislative action requiring the labeling of ingredients in food products is to argue implicitly that government has a role in creating healthy eating habits. Implicit arguments are usually not intended to deceive. They are simply the unacknowledged structures of thought underlying a position. Position takers should be aware of their implicit arguments as well as their explicit arguments. They should critically analyze positions other than their own for implicit arguments, as well.

Why argue? In policy work, you argue to disclose what you think and what you want to accomplish. You do not argue to prove or disprove; you do not argue only for or against. The popular notion of argument as a quarrel between proponents and opponents distorts argument's function and significance for policy purposes. The academic notion of argument as proof also is inadequate for policy purposes. What does argument do in policy activity?

For individual positions, argument displays the underlying reasoning. In public deliberation, arguments disclose the universe of definitions of the problem. In practical politics, argument reveals commonalities and conflicts among positions. These are the grounds on which a course of action can be deliberated. Commonalities among arguments can point to potential cooperation, perhaps compromise, co-sponsorship, or coalition forming. Conflicts among arguments can give insight into competing interests and values that must be taken into account in negotiating a solution.

To illustrate, a farmer has applied to local government for a permit to operate a large-scale industrial farm called a confined animal feeding operation (CAFO). In the rural municipality where the farmer lives, the zoning ordinance allows such operations only as a "conditional" use of land zoned for agricultural uses. "Conditional" uses require case-by-case decisions by local officials on whether to permit or not permit the use. The decision process includes a public hearing inviting residents and others to comment on the proposed use. In the hearing held on the farmer's application, arguments including the following are made:

- Farmers have rights to use and to benefit from their property; to deny this permit is to violate the farmer's private property rights.
- Nearby homeowners have rights to use and to benefit from their property; to grant this permit is to violate the neighbors' private rights.
- Large-scale confined animal farming pollutes the environment and creates human health risks; to grant this permit is to fail to protect natural resources and public welfare.
- Large-scale confined animal farming is regulated and better monitored for compliance with antipollution control than unregulated small-scale farming; to grant this permit will not harm local water or community health.
- Farming is an endangered occupation; to grant this permit will enable a local family-owned farm to succeed by expanding operations and will help to preserve farming in the region.
- Farming is an endangered occupation, and industrial farming is driving smaller farmers out of business; to grant this permit is to harm the local economy, which is still based on diverse types of farming.

If you were a local official, how would you decide this request? Clearly, many arguable issues and competing positions are involved. You might permit or not permit the use, basing your decision (either way) on a single argument. Alternatively, you might focus on a commonality among the arguments such as the wish to preserve rights or the wish to protect public health. Then you might ask the farmer and the neighbors to work out a compromise application. You might delay your decision until you have a revised application that takes specified risks to the community into account.

Argument has its limits in practical policy work, of course. "Arguments are made by all players all the time; as a result they have limited effectiveness. Although arguments are a necessary ingredient to any strategy, they never work by themselves" (Coplin and O'Leary 107). As the local government illustration suggests, you might need to craft a political compromise along with arguing your position. Also, you must recognize political conditions that will determine your argument's effectiveness. Majority control in a governing body has more to do with the acceptability of arguments than does the quality of the arguments themselves. In the local government illustration, the official who represents majority political power might have sufficient influence to force a compromise. The minority power representative might not, unless others can be persuaded to join the minority's position.

When is argument important in a policy process? Argument can make a difference at several points. Arguments matter before a policy process begins, as positions are being developed. They matter at the outset of a process, as stakes are declared and agendas set. They matter again at the end of the process, when a decision is being made.

How to Argue in a Position Paper

Goal: Critical awareness of your own position, critical understanding of other positions, and willingness to consider and to engage other positions.

Objective: Reasoned argument for a position showing awareness of alternative positions and reasoning.

Product: Written document that explicitly argues and aims to persuade.

Scope: Position paper offering either a "big picture" of conditions, causes, or consequences relating to a problem or a "little picture" of significant particulars. As a document, a position paper might run to book length in some circumstances; typically it is briefer, two to three pages.

Strategy: Your position in relation to others:
- Make a list of the known positions on the problem.
- Ask and answer the questions "What does my position have in common with others on this list?" and "How does my position differ from or conflict with others on the list?"
- Note specific commonalities, differences, and conflicts of values, assumptions, or ideas between your position and other positions.
- Identify potential grounds for cooperation and for competition.

Task #1. Outline your argument.

If you are authoring a position paper for a professional association or for a nonprofit organization, make sure you understand its mission and how the position you are taking relates to the mission. Be clear on that relationship. Consult before deviating from the mission.

In most cases, you can use the following outline for informal arguments to construct the logic of a policy position:

- Problem
- Issue
- Question about the issue that has at least two answers and is therefore arguable
- Claim (the arguer's assertion or answer to the question)
- Support:
  - Justification
    1. reasons ("because," or the relevance of the assertion)
    2. assumptions ("why," or the values, beliefs, and principles that motivate the assertion as well as the authority represented in the assertion)
  - Elaboration
    1. grounds (supporting evidence for the reasons and the assumptions)
    2. limits (constraints the arguer would place on a claim)
  - Anticipated reactions (potential responses from diverse other positions)
    1. Cooperative or supporting assertions
    2. Competitive or opposing assertions
    3. Altogether different assumptions
    4. Challenges to reasons or to grounds

Note: The outline does not include rebuttal. A position paper should not rebut. Rather, it should state its reasoning in a way that shows that reactions to the reasoning have been anticipated.

Task #2. Write the position paper.

Review the method in chapter 2 before you write.

By consciously thinking about your position (see Strategy, above) and outlining the logic of your argument, you have already begun to plan the contents of the document. That does *not* mean that the document's contents should simply fill in the outline. In the document, information should be organized according to the needs of readers as well as the aims of the writer.

An outline that helps a writer to think might not help readers to understand.

The message that the document conveys typically will be your claim or your answer to the issue question. Or it will be a conclusion that you draw that derives from the claim.

You must clearly understand your authority and make it clear to the readers. Your authority for taking a position derives from your role and power in the process as well as your credentials.

The document must clearly show whose position it communicates. Yours? That of an organization that you represent? Then you must anticipate reactions to your position. Go back to the list you made of positions other than your own (see Strategy, above). To each position on the list, add the reaction you might accordingly expect, and then rank the reactions in order of importance to you.

Anticipate responses, but do not rebut them in the position paper (unless you are directed to do so). Keep the focus on your position.

Condense greatly, for now. You will likely have later opportunity to elaborate. However, keep this in mind: Ignoring information your readers may ultimately decide (under the influence of other arguments) is important will cost you credibility. By writing your position paper to show that you are aware of both similar and conflicting views, you retain credibility.

Put lengthy evidence in an appendix. Charts, tables, other graphics, or extended textual materials should be appended. However, the choice to append important details should rest on knowing the circumstances in which the position paper will be read and used. Writers especially should know whether all readers will see the entire document, including appendices.

Use a standard citation style for identifying sources. Modern Language Association (MLA) style might be sufficient. Use government style (see chapter 3) if you are citing government records.

If you are authoring a position paper that speaks for a group or organization, plan to allow adequate time for consultation. Are you the sole author, or do you have collaborators? Are you ghostwriting for someone else? Who will review drafts and make revisions?

Remember to check the final product against the standard (see checklists, chapter 2).

Example

This position paper was written in memo form by the student shown in Scenario 2 at the beginning of this chapter. She produced the position paper in a writing course based on analysis she had performed in a public policy studies course.

THE DOMESTIC WAR ON DRUGS AND THE HIGHER EDUCATION AMENDMENTS OF 1998

Memo

To: John Walters, Director of the Executive Office of National Drug Control Policy
From: Sandra Derstine, Representative of National Students for Sensible Drug Policy
Date: October 23, 2001
Re: Higher Education Act (HEA) Amendments of 1998

Executive Summary

In recent decades, the domestic war on drugs has grown as a primary target of legislative concern, as well as popular attention. In 1998, amendments to the Higher Education Act of 1965 were signed into law. One of the provisions in this law restricts federal financial aid to any student convicted of a prior drug-related offense, no matter how minor. Supporters of this law cite it as a major victory over drug abuse, exemplifying the zero tolerance policy of the United States. Opposition, who undoubtedly would also like to see drug abuse decline, disagree with the merit of this path. Those who would like to see the amendment altered see its flawed logic and many gaps—it disproportionately affects working class and minority students, tries students twice for the same crime, and does not appropriately support the drug treatment programs that it proposes. In order to solve the drug problems of this country, education should be encouraged, not denied.

What Is the Higher Education Act? What Are the HEA Amendments of 1998?

The Higher Education Act (HEA) was created in 1965 to open higher education opportunities for all Americans. It establishes federal financial aid programs such as Perkins Loans, Pell Grants, Supplemental Educational Opportunity Grants, PLUS Loans, and Work-Study Programs. The Act is periodically reviewed and updated by Congress to ensure adequate funding and access to college for millions of Americans.[1]

In 1998 the HEA was amended to include a new provision that closes college opportunities to students revealing drug convictions on their application forms. The amendment restricts federal financial aid eligibility to all students with any prior drug-related convictions, no matter how minor.[2]

Why Should the HEA Amendments of 1998 Be Changed?

1. *Restricting eligibility for federal financial aid hurts working class families.*

Since federal aid is need-based, only drug-convicted students of working and lower class families will be disproportionately affected by the new amendments. Upper class families with drug-convicted children can afford both a good lawyer and a good education and therefore will never need to disclose the information on a federal financial aid form or be affected by the new policy.[3]

2. *Restricting eligibility for federal financial aid is discriminatory.*

In New York State, almost 95% of those in prison for drug offenses are people of color, but the majority of people and the majority of drug users are white.[4] African Americans, who comprise approximately 13% of the population and 13% of all drug users, account for more than 55% of those convicted for drug offenses. There is no reason to believe that the disproportionate racial impact of drug law enforcement won't spread into the realm of higher education via this law.[5]

3. *This amendment tries students twice for the same crime*. The students who are having their aid restricted have already paid the price the criminal justice system demands upon their conviction (whether it be serving time, probation, or community service). It makes no sense to punish students again by limiting their ability to get an education and improve their lives. Additionally, judges handling drug cases already had the option (prior to the implementation of this law) of denying drug offenders federal benefits. School administrators also have the power to expel problem students. These are the people who know the students best and should be the ones who decide their educational futures—not the federal government.[6]

4. *This amendment does not support the drug abuse treatment that it proposes*.

This law allows federal financial aid to be reinstated for students once they have successfully completed a qualifying drug-treatment program. Unfortunately though, the law does not increase appropriations for this provision. Treatment accounts for only 15% of the drug budget; hence, most of the people who need it are not getting it. And those who cannot afford an education without financial aid probably cannot afford a private treatment program or the time off from work or school it takes to complete one.[7] A recent study by researchers at Substance Abuse Mental Health Services Administration (SAMHSA) has indicated that 48% of the need for drug treatment, not including alcohol abuse, is unmet in the United States.[8]

What Arguments Support the HEA Amendments of 1998?

1. *Students who have been convicted of drug-related offenses are bound to do it again; therefore we should not waste federal money on them if their cause is futile*.

This is very misinformed logic. This assumes that people do not change. In fact, adolescents have the most maturing and changing to do, and an education is the best way to ensure that an adolescent moves toward improving a past lifestyle.

2. *The Gateway Theory, which exemplifies marijuana as a drug that begins a person's descent into uncontrollable addiction of harder drugs, leads to the conclusion that users of any drug, no matter how minor, should be treated the same. In this case, the logic is that all students should be denied aid because they are all bound to do hard drugs in the end.*

The Gateway Theory is false. In March 1999, the Institute of Medicine stated, "There is no conclusive evidence that the drug effects of marijuana are causally linked to the subsequent abuse of other illicit drugs." In fact, over 72 million Americans have used marijuana, yet for every 120 people who have ever tried marijuana, there is only one active, regular user of cocaine.[9]

3. *The Amendment only restricts eligibility until a student's successful completion of a qualifying drug rehabilitation program.*

This argument is true. Through the HEA Amendments of 1998, federal financial aid eligibility is only permanently restricted for those who have not completed a treatment program. But, the flaw in this provision is that it establishes a high standard, yet does not authorize any new spending. It fails to anticipate possible appropriations instability due to an increase in student participation in these programs.[10] Therefore, the students that want and need treatment are most likely not going to get it. Drug treatment should be provided on request. Treatment has been shown to be 10 times more cost effective than interdiction in reducing the use of drugs.[11] Studies also reported by the White House Drug Policy Office show that for every $1 spent on treatment, $7 is saved in criminal justice, health care, or welfare costs that would otherwise be borne by society.[12]

Why Should You Support Changing the HEA Amendments of 1998?

This amendment will not solve our drug war. The solution is not to deny our citizens an education, but to give them one. Setbacks such as not being able to raise money for school can set a young ex-offender into a downward spiral toward failure. This surely will not set right the drug problems we have in our country, but instead will proliferate and intensify them.

The solution to our drug problems lies within educational programs. In years past our focus has been misdirected. For example, from 1987 to 1995, state spending on higher education decreased by 18.2%, while spending on corrections increased by 30%. Keeping kids in school and making educational opportunities more available should be our focus. Studies have shown that "alternative programming appears to be the most effective among those youth at greatest risk for substance abuse and related problems." An independent study by Public-Private Partnerships evaluated Big Brother/Big Sister programs and found participants to be 46% less likely to start using illegal drugs and 27% less likely to start using alcohol.[13]

The latest numbers from the Department of Education (9/23/01) indicate that a record number of students are likely to lose their full-year financial aid eligibility this year. For the 2001–2002 school year, 279,044 applicants who refused to answer the drug question had their application processed, and 9,114 had their aid cut after revealing a drug conviction. Tighter enforcement of the existing law is behind the dramatic increase in the number of applicants being denied aid. For the 2002–2003 school year, students who do not answer the question will not have their application processed. It is estimated that between 40,000 and 60,000 students will be formally denied aid for some or all of this school year, plus an unknown number who don't bother to apply because they rightly or wrongly believe they are ineligible.[14] That means that between 40,000 and 60,000 students will be back on the streets in a downward spiral towards failure. And their lost potential will be on the hands of the federal government.

Sources

1. Raise Your Voice. *Frequently Asked Questions About the Higher Education Act Drug Provision.* 20 October 2001. <http://www.raiseyourvoice.com/heainfo. html>

2. U.S. House of Representatives. 105th Congress. 2nd Session (1998). "H.R. 6, Drug Free Student Loans Amendment." Version: 5; Version Date: 1/10/97. (Full Text of Bills: Congressional Universe Online Service. Bethesda, MD: Congressional Information Service.)

3. Raise Your Voice. *Frequently Asked Questions About the Higher Education Act Drug Provision.* 20 October 2001. <http://www.raiseyourvoice.com/heainfo. html>

4. Raise Your Voice. *Frequently Asked Questions About the Higher Education Act Drug Provision.* 20 October 2001. <http://www.raiseyourvoice.com/heainfo. html>

5. The Sentencing Project. *Briefing/Fact Sheets.* 20 October 2001. <http://www.sentencingproject.org/brief/brief.htm>
6. Raise Your Voice. *Frequently Asked Questions About the Higher Education Act Drug Provision.* 20 October 2001. <http://www.raiseyourvoice.com/heainfo.html>
7. Raise Your Voice. *Frequently Asked Questions About the Higher Education Act Drug Provision.* 20 October 2001. <http://www.raiseyourvoice.com/heainfo.html>
8. The Razor Wire. *Alternatives to the Drug War.* 20 October 2001. <http://www.november.org/1101.html>
9. Drug War Facts. *Factbook: Gateway Theory.* 20 October 2001. <http://www.drugwarfacts.org/gatewayt.htm>
10. Hill Source Issue Briefs. *House Republican Conference: A Drug Free America by 2002—Winning the War on Drugs to Protect Our Children.* 20 October 2001. <http://hillsource.house.gov/IssueFocus/IssueBriefs/IBMain/Drugs520.pdf>
11. The Razor Wire. *Alternatives to the Drug War.* 20 October 2001. <http://www.november.org/1101.html>
12. Raise Your Voice. *Frequently Asked Questions About the Higher Education Act Drug Provision.* 20 October 2001. <http://www.raiseyourvoice.com/heainfo.html>
13. The Razor Wire. *Alternatives to the Drug War.* 20 October 2001. <http://www.november.org/1101.html>
14. Raise Your Voice. *Frequently Asked Questions About the Higher Education Act Drug Provision.* 20 October 2001. http://www.raiseyourvoice.com/heainfo.html

What This Example Shows. This credible and concisely worded position paper makes clear its authority (the writer's role, power, and representation) and its message (the author's position) (see method, chapter 2). It is also well argued. It lays out the logic of the author's position, fairly summarizes other positions, and anticipates reaction (see Task #1, this chapter).

Too much attention goes to rebutting the opposition. Save rebuttal for occasions when debate is the purpose. In a position paper, while acknowledging other positions, you should focus on laying out the reasoning that supports your position.

The memo format with overview and subheadings makes this position paper readable. However, a word about its presentation of evidence: Caution should be taken when evidence is mostly quantitative. Caution is especially warranted in policy contexts because many relevant and persuasive data in policy argumentation are quantitative (see checklists, chapter 2).

Here's why caution is wise: From the perspective of communication in a public forum, numbers are harder to comprehend than words. Depending on the audience, some readers (and especially listeners) might have trouble absorbing quantitative information. Trouble is most likely to occur when raw numbers, statistics, dates, or dollar amounts are clustered together. This position paper includes several instances of risky clustering. The paper's overall usefulness would be improved if the writer paraphrased or used suitable graphics for some (not all) of the quantitative information, as illustrated below.

Original

For example, from 1987 to 1995, state spending on higher education decreased by 18.2 percent while spending on corrections increased by 30 percent. . . . An independent study by Public-Private Partnerships evaluated Big Brother/Big Sister programs and found participants to be 46 percent less likely to start using illegal drugs and 27 percent less likely to start using alcohol.

Possible revision

For example, state spending on higher education decreased while spending on corrections increased. [Follow this sentence with a graphic showing dates and spending trends. This is a strategy of separating claims from supporting data and of rendering quantitative data graphically.] . . . An independent study by Public-Private Partnerships evaluated Big Brother/Big Sister programs and found participants to be 46 percent less likely to start using illegal drugs and 27 percent less likely to start using alcohol. [Keep these statistics. This is a strategy of selective use, allowable if distracting overuse has been reduced.]

Note: Choice of symbols for representing quantitative information might be left up to the writer or might be prescribed by a house style. The chosen style should be used consistently; do not mix "46%" and "forty-six percent" in a document.

Reference

Coplin, William D., and Michael K. O'Leary. *Public Policy Skills*. 3rd ed. Washington, D.C.: Policy Studies Associates, 1998.

❖ 6 ❖

PETITIONS AND PROPOSALS:
REQUEST ACTION
OR PROPOSE POLICY

Scenario 1

An undergraduate student is a longtime volunteer in a local women's center that participates in a statewide coalition of private and public groups concerned about domestic violence. As a volunteer supporting the center's small administrative staff, the student is often given writing tasks. She has produced public information documents including brochures, guides to services, and how-to instructions. This time, her supervisor asks her to draft a policy statement to communicate the center's advocacy for amending proposed legislation regarding guns and domestic abuse. The center director will present the statement in a public hearing on the legislation.

Scenario 2

An undergraduate student who is majoring in both public policy studies and education plans to be a teacher. She supports a current reform, charter schools, whereby administrators, teachers, and parents may design public schools around a chosen academic focus and targeted learning objectives. Having researched legislative authorization for charter schools at the federal and state levels, she recognizes a gap in the plans for implementing the reform. The gap is accountability. She takes advantage of a classroom simulation, a roundtable of nongovernmental advocacy organizations, to make a policy proposal to address the accountability gap.

Scenario 3

Two professionals in auto safety create a for-profit corporation in the public interest in order to assist victims in litigation resulting from automobile accidents. Both are experts, one of them in highway safety policy and the other in automobile design for occupant safety. Concerned about the high rates of injury and fatality caused in part by drivers' and passengers' failure to use seat belts, the experts petition the responsible federal government agency to amend the safety standard to require inducements in all vehicles to encourage seat belt use.

Nongovernmental groups, as well as individuals, may request government action or propose public policy. This chapter shows you how to propose policy on behalf of a group.

In the United States, at the federal level and in many states, only elected legislators are authorized to originate and enact laws. However, requests for action and policy proposals may originate inside or outside government. This chapter focuses on external requests, particularly petitions and proposals to government for legislative or administrative action.

One longstanding practice is petitioning. The First Amendment to the U.S. Constitution guarantees citizens' right to "petition their national legislature for a redress of grievances." Over time, petitioners have come to include not only individual citizens but also groups, organizations, and corporations of many kinds. Petitioning has extended beyond redressing grievances to requesting varied actions.

To illustrate petitioning, in a case of injury due to air bags deployed in automobile collisions, three different petitions for government action might be made:

1. A victim might petition his congressional representatives to amend the National Traffic and Motor Vehicle Safety Act of 1966 to authorize training programs for emergency services personnel.
2. A company that has developed a new technology for increasing passenger safety without relying solely on passenger restraints such as air bags or safety belts might petition the National Highway Traffic and Safety Agency to test the new technology.
3. A professional association of automotive engineers might petition the National Highway Traffic and Safety Agency to amend a vehicle safety design standard to include warning systems in cars to encourage seat belt use.

The other common practice is proposing policy. Proposals usually represent organized, or group, interest in solving a problem. The role of nongovernmental groups in North American public policy making has deep historical roots. In colonial America, before the United States or its government was established, voluntary associations flourished. Individuals formed associations to provide basic social services, to meet public needs, and to protect community interests. Voluntary fire companies, water companies, library associations, prison associations, school associations, landowner associations, and militias were so common in the America of the early 1800s that a visitor from France, Alexis de Tocqueville, observed, "Americans of all ages, all conditions, and all disposition constantly form associations. . . . Wherever at the head of some new undertaking you see the government in France, or a man of rank in England, in the United States you will be sure to find an association" (106). Such group activism provides background for the Fifth Amendment regarding limitations on central government that states "the powers not delegated to the United States by the Constitution, nor prohibited to it by the States, are reserved to the States respectively, or to the people."

Nowadays, groups that perform a public good might be granted tax-exempt status as nonprofit organizations. Their function might be religious, scientific, literary, educational, promotional, protective, political, charitable, or other, in accordance with Internal Revenue Service standards for twenty categories of tax-exempt activity (Internal Revenue Service website). In 1998, over a million nonprofit organizations operated in the United States (*The New Nonprofit Almanac IN BRIEF*).

Many (possibly most) nonprofit organizations are not concerned with public policy. However, significant numbers of such groups are actively involved in attempting to influence the direction public policy takes. These are known as advocacy groups. Their methods vary according to the limitations of their tax-exempt status. Some limit their activity to education. These educate their members, the larger public, and the government regarding issues, but they do not lobby legislators or support candidates for election. Their communications include legislative alerts, editorials and letters, personal visits to lawmakers, witness testimony, and more. Others, with more restricted tax benefits, might campaign for candidates running for office or lobby for outcomes of a process or provide lawmakers with expert information and political assessments as well as, on occasion, drafts of legislation.

Legislators often appreciate the help of advocacy groups in educating the public about needs for policy. Government staffs appreciate informed, accurate, well-argued lobbying because it helps them to brief legislators on complex or controversial issues. A positive example of public good resulting from such help might be the continued strengthening of legislation in the United States on smoking as a health problem. Past legislation on smoking has been passed in large part because health care advocacy groups worked with responsive legislators at all levels of government and educated the public to support directed warning labels on cigarettes, nonsmoking restaurant sections, and smoke-free public facilities. A negative example is the influence of lobbying by corporations and advocacy groups to weaken laws on occupational health and safety or on environmental protection.

Grassroots organizations such as neighborhood or block associations, community clubs, workplace voluntary groups, and student organizations might also use petitioning or proposing to accomplish their advocacy, just as nonprofit organizations might.

Why are petitioning and proposing important? They sustain democracy; they are democratic ways of addressing public problems by institutional means. Whether by direct democracy (as with California's state referenda) or by representative democracy (as with Washington's federal legislation), self-governing society relies on procedures for public intervention in the process. Recall that public policy has far-reaching effects in the everyday life of society. Policy makers need and want information that can solve problems and build public support for action. Nongovernmental groups or individuals who are informed about the impact of a problem or a policy are excellent

sources of information. So are groups or individuals who recognize a
need for policy.

Who petitions and proposes? Individuals can do so, but petitions
and proposals by organized groups are likely to have more influence.

How to Ask for Action or Propose Policy on Behalf of a Group

Goal: Knowledge of the functions of nongovernmental organizations in public policy processes, and familiarity with nonprofit organizations active in your area of interest.

Objective: Petitioning and proposing on behalf of an organization or group.

Product: Brief written policy proposal representing an organization's advocacy. Length varies according to purposes and situations, but a short proposal (one to three pages) is preferred.

Scope: Content of group's charter, purpose, or mission to determine the concerns or issues you will address.

Strategy: Proposals with this information:

- Desired outcome: What do you want to accomplish? Can you describe it as if it were already accomplished in a future you want to achieve?
- Today's situation: What's wrong in the present? Why is the action you propose needed? What causes the need?
- Relevant background: How did the problem arise? What original assumptions are no longer valid? What conditions have changed?
- Available options: What are the alternative ways of meeting the need? Advantages and disadvantages of each? Costs (money, other) of each?
- Recommended action: What is the best alternative? Can you briefly argue as to why?
- Summary: What are the results (referring to the desired future) if requested action is performed?
- Action items: Who is asked to do what, when, where, and how?

Task #1. Name the need, and specify the action and agency.

Step one: Identify a need for policy.   If you already know the
need you will address or the option you will advocate, proceed

to the next step, specifying the desired action and the responsible agency.

If you do not know the need or have not decided on an option, or if you are responsible for selecting among many competing needs or options, step back to focus before you proceed. Start wherever you need to start, whether it is to define the problem and pinpoint the issue (discovery), review the history of action or inaction (legislative history), review the arguments (the range of positions), or use the method in chapter 2 to reconsider the policy context as well as the communication situation for your proposal.

Step two: Specify the action and agency. Determining the needed action—knowing what is possible, knowing whom to ask, and knowing what to ask for—is not simple. Much time and effort can be wasted in seeking unlikely action or making a proposal to the wrong recipient. The best way is to continually and iteratively ask and answer, "What am I trying to do?" and "How can I do it most effectively?" That will lead to further guiding questions such as "Should I start small (that is, local), or should I start big (national or international)?"

Consider the options for action; for example, choose government action. What do you want government to do? Government can legislate, spend, regulate, and enforce, all within limits. Which type of action is needed for the problem you are concerned about? To which level of government—federal, state, local—should you direct your proposal? Which department or agency can do what you want to accomplish?

Now consider nongovernmental options. Does the solution require government action at all? For example, a citizens group might choose to organize a boycott or initiate a lawsuit to solve a civic problem rather than to ask for government action or propose public policy. Similarly, a student group might choose a community solution rather than governmental action. For example, in response to a racist incident on campus, one student group developed a constructive plan for educating students about everyday racism in campus life. Rather than proposing it as policy to student governance or to the school's administration, the group circulated their plan among other campus organizations and sent it to national student associations. They communicated it by word of mouth and publicized it through

news media. The strategy was to ask similar student groups na-
tionwide to draw public attention to the problem of race-based
harassment on their campus and to offer as a model the original
group's plan for addressing it. In this example, change in human
behavior was sought through organized community education.

Task #2. Identify the organizations active on your issue.

Here are some ways of locating and identifying nonprofit
organizations:

- Check the local phone directory, or ask local volunteer
  services about local nonprofits or local affiliates of national and international nonprofits.
- Ask a librarian for national guides to nonprofit
  organizations.
- Read the transcripts of congressional hearings on your
  issue to find witnesses who spoke on behalf of advocacy
  groups (see chapter 3).
- Search newspaper databases for articles on your issue
  that might refer to advocacy groups.
- Search World Wide Web portals to nonprofit organ-
  izations. Some offer free searchable lists of nonprofits
  such as:
  - *Institute for Nonprofit Management:* http://inom.org
  - *Nonprofit Online News:* http://news.gilbert.org
  - *Nonprofit Nuts & Bolts:* http://www.nutsbolts.com
  - *Internet Nonprofit Center:* http://www.nonprofits.org
  - *Minnesota Council of Nonprofits Find a Nonprofit:*
    http://www.mncn.org/find.htm
  - *idealist.org:* http://www.idealist.org
  - *Independent Sector:* http://www.independentsector.org
  - *Nonprofit Pathfinder:* http://www.indepsec.org/
    pathfinder/index.html
- Try these subscription services for details including tax-
  exempt status and financial information on specific non-
  profits:
  - *Associations Unlimited:* http://galenet.galegroup.com/
    servlet/AU?locID=syra96044
  - *The Foundation Directory Online:* http://lnps.fdn
    center.org
  - *Guidestar:* http://www.guidestar.org

Why restrict your knowledge of the players to nonprofit organizations? If, for example, you represent a health care organization and are advocating for the right to use a controversial drug, you may want to enlist the support of the pharmaceutical company that manufactures the drug. While the company has a vested interest, it also might have facts and figures that might bolster or undercut your arguments.

Task #3. Write a policy proposal.

Absolutely key in petitioning and in proposing is providing only accurate information. Anything else will destroy your or your organization's credibility. Use the method in chapter 2 to prepare, plan, and produce a written proposal. The document's contents should answer the questions listed under Strategy (this chapter, above). Compare the finished product to the standard (see checklists, chapter 2).

There is no typical format for policy proposals. As noted elsewhere in this guide, if you are writing for an organization that prescribes a template for policy proposals, use that template. Ordinarily, the conventions of professional communication will apply. Use a header that provides identifying information, an overview that summarizes the proposal, and subheaded subsections that provide information. The document type might be a letter, a memo, a full-page ad in a national newspaper, a public declaration dramatically delivered in historical costume, or another form chosen for its effectiveness in the situation. See, for example, the websites of national nonprofit groups that sometimes express their advocacy in funny as well as serious ways and in attention-grabbing modes.

Three Examples

Example 1. Here is a policy proposal by a nonprofit organization advocating an amendment to proposed legislation. The student author, who is shown in Scenario 1 at the beginning of this chapter, wrote the proposal for a course assignment in public policy writing. She used her experience as a volunteer ghostwriter for the organization's spokesperson to recreate a proposal for the assignment.

POLICY PROPOSAL

Who I Represent

For the past 20 years the Maryland Network Against Domestic Violence (MNADV) has been working to end domestic violence against
women. MNADV works with domestic violence service providers
and criminal justice personnel throughout the state to provide consistent community responses to domestic violence. In support of
community response, MNADV focuses its lobbying efforts on
changes needed in state law and has aided in passing almost thirty
pieces of domestic violence legislation. Currently, we support the
passage of HB146 Domestic Violence Protective Order Additional
Relief. This legislation provides for legal procedures requiring domestic abusers to surrender firearms after a protective order hearing.

Our Position

The issue is whether a person who has been accused of an act of
domestic violence should be allowed to own firearms. As a representative for MNADV, I am here to say that we strongly feel that
keeping guns out of the hands of batterers will help prevent further
physical injury to victims. In Maryland, the majority of domestic violence incidents involve a gun or other firearm.

Currently, Maryland has no law that keeps guns away from
batterers. As a result, men who are convicted of domestic violence
often end up going back to their victims, with a gun, when their sentence is over. If Maryland does not pass a law preventing batterers
from owning firearms, victims who have survived are more likely to
become victims again. If Maryland fails to pass such a law, the state
is also failing to adequately protect abused women.

If a person convicted of or accused of domestic violence is
charged with a misdemeanor for owning a firearm as HB 146 provides, the number of deaths due to guns in the case of domestic violence will be lowered.

Example 2. As described in Scenario 2, a student wrote and presented this proposal as the spokesperson for a nonprofit organization in a (simulated) roundtable held annually by a coalition of nonprofit organizations to set the coalition's lobbying agenda for the year.

MEMORANDUM

To: Umbrella Action
From: The Center for Education Reform
Date: March 28, 2002
Issue: National Charter School Evaluation

Problem Statement

Thirty-seven states have unique charter school legislation. Federal legislation has recently granted the Secretary of Education more power to allocate increased financial assistance to state educational agencies (SEAs) for the authorization, implementation, and maintenance of charter schools. Assessment, however, remains one of the most variable components of such legislation. Evaluation methods are often ambiguous to both state and local education agencies (LEAs). In addition, diverse policies make it nearly impossible to draw accurate conclusions about the status or progress of charter schools as a national educational reform. The National Education Association reports that "systems for accountability in charters have been inconsistent and ill-defined." Many non-profit and government agencies have similar concerns and study charter school data with little consistency. It is necessary to take a new approach to evaluating charter schools and the Center for Education Reform is prepared to be aggressive in this approach.

Proposed Solution

There are currently no public policy proposals to address this problem. Three alternatives should be considered. One means to ameliorate the problem is to create a governing body to create a set of assessment guidelines that all state charter school legislation will include. Another solution is to stop trying to evaluate and report national charter school data. Finally, creating an all-encompassing federal charter school law would be another way to improve assessment.

The Center for Education Reform requests your support in advocating for a regulatory board within the Department of Education to be established for the annual assessment of charter schools. This body, The Council for Charter School Evaluation, will create basic standards which SEAs and LEAs must require of all charter schools. These standards will be used to compare and assess schools across the nation, determine their success and progress, and measure the impact charters have on participating individuals and groups. The U.S. Department of Education needs one year of planning before the council is ready for operation, necessitating the immediate adoption of this policy. The implementation of this policy initially requires the following actions.

The United States Department of Education must allocate at least 1% of President Bush's proposed $56.5 billion education budget for 2003 to establish this council. President Bush and Secretary of Education Roderick Paige must select members of the council after member characteristics and requirements are determined. The council members must spend time examining current charter school data and state legislation, listening to experts on continuous improvement and assessment, and determining standard criteria and evaluation processes.

Cost/Benefit Analysis

The benefits of this policy are substantial. Charter school authorizers will initially know how, when, and by what measures their school will be assessed and compared. Failing charter schools will not be permitted to stay open. Non-profit organizations and government agencies will be able to make more accurate conclusions from data the counsel provides. Charter schools will also be taken more seriously as national education reform. Finally, students, parents, educators, and communities will be able to be more secure in their investment in charter schools.

Our policy, however, is not without costs. Money allocated from the education budget will reduce financial assistance to other programs. There will be more work for pre-existing charters to meet the new standards. Our proposed policy is more politically feasible than the alternatives because it is less extreme and charter schools receive much bipartisan support. However, our innovative recommendation illustrates the conflicted argument between more or less government involvement in American institutions. The political debate will be a cost to the policy because there will be less support than opposition. Establishing the Counsel for Charter School Evaluation is a very effective way to improve the way charter schools are assessed. We will, however, need the support of Umbrella Action in order to get this policy to the mainstream table educational of reform debate.

Example 3.   Here is a petition to a government agency by the auto safety experts described in Scenario 3. In a letter of transmittal, they addressed the petition to the head of the responsible agency.

LETTER

Carl E. Nash
(address)
(telephone number)
(e-mail address)

December 17, 1998

The Honorable Ricardo Martinez, M.D., Administrator
National Highway Traffic Safety Administrator
400 Seventh Street, SW
Washington, D.C. 20590

Dear Dr. Martinez:

Enclosed is a petition for the National Highway Traffic Safety Administration to amend Federal motor vehicle safety standard 208 to require an effective safety belt inducement in all new motor vehicles. It has been a quarter century since the unfortunate experience with safety belt ignition interlocks. You have an obligation to seriously reconsider the potential of vehicle-based systems to substantially increase belt use.

The advantages of an acceptable, effective belt use inducement are substantial. It would reduce fatalities by at least 7,000 per year and would reduce injuries comparably. It would permit the agency to respond favorably to the industry's desire that NHTSA rescind the unbelted test in FMVSS 208. It would end the controversy over the use of safety belt use laws as an excuse for stopping minority drivers.

Donald Friedman and I have also submitted comments to the docket of the advanced air bag rulemaking notice that build on the concept of an effective safety belt use inducement. We believe that a simpler and more effective approach to reducing inflation induced injuries can be based on this concept.

In the interest of advancing motor vehicle safety, we look forward to your favorable consideration of our petition and of our comments on the rulemaking proposal. Action that could make belt use nearly universal in the United States is long overdue, and would be an important legacy of your tenure.

Sincerely,
Carl E. Nash, Ph.D.

PETITION

To Amend FMVSS 208, Occupant Crash Protection
To Require Effective Belt Use Inducement

Carl E. Nash, Ph.D.

Donald Friedman
Washington, D.C., and Santa Barbara, California

Summary

This is a petition to amend Federal motor vehicle safety standard 208 (FMVSS 208) to require effective safety belt use inducement systems in all new motor vehicles sold in the United States. This requirement should become effective no later than the beginning of the 2001 model year. The inducement systems should activate only if a person sits in either front outboard seating position and does not attach the safety belt after occupying the seat and would stop when the belt is buckled. The requirement must be consistent with the 'interlock' amendment to the National Traffic and Motor Vehicle Safety Act of 1966 (15 U.S.C. 1410b), which prohibits ignition interlocks and continuous buzzers.

The inducements could include, but need not be limited to: (1) a continuous visual warning to buckle safety belts located prominently on the instrument panel, (2) an intermittent, repeating audible suggestion (such as with a synthesized voice) warning occupants to buckle their safety belt, and (3) disruption of electrical power to such non-essential accessories as the radio, tape or CD player, and air conditioning. We further recommend that NHTSA undertake a quick reaction project to determine the acceptability and effectiveness of various types of use inducements to ensure that the spirit of the interlock amendment is not violated.

Background

[Detail not shown here.]

Restraint Policy and Use Today

[Detail not shown here.]

An Amendment to FMVSS 208

Therefore, we petition NHTSA to amend FMVSS 208 to require a reasonable and effective safety belt use inducement to be built into all new vehicles. Effective belt use inducements can be required without violating the 'interlock' amendment (15 U.S.C. 1410b) to the National Traffic and Motor Vehicle Safety Act [detail omitted].

Safety belt use is widespread, generally accepted, and required by law in virtually all states. The design for comfort and convenience of safety belts in many new vehicles has improved since the days of the interlock. Thus, we doubt that many motorists would object to use of well-designed inducement systems. However, we recommend that NHTSA conduct quick reaction tests using panels and field tests

to determine effectiveness and consumer acceptance of various types of use inducements.
[Detail omitted.]

We note that a policy of increasing belt use through an inducement built into new motor vehicles would be preferable to the present policy of safety belt use laws for reasons unrelated to safety. Civil rights organizations (most recently the Urban League) have objected to primary belt use laws because of their potential to give police officers an excuse to stop minority drivers. Having the inducement built into the vehicle takes away that issue and should be strongly supported by civil rights and civil liberties advocates.

Requiring a belt use inducement built into all new vehicles would be a major improvement in every way to FMVSS 208. As existing cars are retired from use, it would increase belt use to near universality (with the attendant reduction in fatalities and serious injuries in all crash modes) without further state laws or enforcement activities. In fact, states could sunset their safety belt use laws within the next decade or two. We estimate that a belt use inducement has the potential to save a minimum of 7,000 additional lives per year.

We urge that NHTSA give priority to both testing and simultaneous rulemaking in response to our petition.

(The full petition can be found in NHTSA 98-4405-62. To find it, at dms.dot.gov, select simple search and type in 4405 as the docket number. The comment is #62. The agency's response denying the petition can be found in Appendix A: Response to Petition at http://www.nhtsa.dot.gov/cars/rules/rulings/AAirBagSNPRM/AppA.html.)

What These Examples Show.   Example 1 illustrates a pared-down, to-the-point presentation of information. Because oral delivery is intended, the advocacy statement is brief. It includes only the message and key evidence. In a briefing or a public hearing, time limitations usually force the omission of details. Details can come out in later question-and-answer sessions. They are also available in the full written statement that is usually provided to the agency or committee holding the hearing. Additionally, a staff member for the hearing's convener might follow up afterwards by asking the organization's spokesperson for more information.

Example 1 proposes policy. Its chosen genre is advocacy statement, in oral and written versions. Example 2 analyzes policy options and recommends an alternative. Its genre is a policy analysis memo. Example 3 requests a change in policy. Its genre is the petition. All three examples are situated in a specified policy process (see method, chapter 2).

Examples 2 and 3 illustrate in-depth presentation of information. They are written to be read silently, not spoken aloud. (One cue to this intention is the length of sentences in both texts. Many sentences in Examples 2 and 3 would cause a speaker to run out of breath. Another cue is the amount of details.)

Depending on the communication situation, the amount of details in Examples 2 and 3 might be appropriate. For both petitions and proposals, the contents might be prescribed by the receiving agency; length might also be prescribed. If the agency prescribes the information it wants and the length or organization it prefers, writers should provide these as prescribed. Otherwise, the petition or proposal might go unconsidered (see method, chapter 2).

Where to put all the details? Putting them in the main body of the document can be an unwise choice. Detailed explication or historical background presented alongside your message can bury the message. Example 2, the policy analysis memo, buries its message in this way.

Example 3, the petition to amend policy, is well organized as a whole. Its structure aids rapid comprehension. A summary states the message and a condensed version of supporting information. This summary and later subheadings focus attention on the message while developing the support. Even so, the message could be emphasized more. Within paragraphs, the petitioners' implicit message ("There is a better way") competes with the implied critique ("The agency's way is flawed"). Impatient readers might want a sharper focus on the message.

A choice for writers of Examples 2 and 3 is to put explanation, background, and critique in an appendix. The success of this option depends on readers' circumstances. To decide whether to append supporting information, writers need to understand the situation in which their documents will be read. If all readers will receive the whole document and if they are willing to flip between the main text and appendix as they read, writers can safely choose to append detailed information. However, in some settings of policy work, readers are likely to receive not the whole document but only the parts that pertain to their jurisdiction or responsibility. Appended information might not reach them.

All three examples include information and are credible. Each makes sufficient disclosure of evidence, organizes evidence persuasively, and provides for accountability by identifying the presenter and the recipient (see checklists, chapter 2). The credibility of Example 2 is weakened, however, by including a quotation for which no source is cited. The quotation cannot be traced and verified.

References

Internal Revenue Service. *The Digital Daily.* 2 May 2002. <http://www.irs.gov>.

The New Nonprofit Almanac IN BRIEF: Facts and Figures on the Independent Sector. 2 May 2002. <http://www.independentsector.org>.

Tocqueville, Alexis de. *Democracy in America.* Vol. 2. 1840. New York: Alfred A. Knopf, 1945.

❖ 7 ❖

BRIEFING MEMO
OR OPINION STATEMENT:
INFORM POLICY MAKERS

Scenario I

The date is September 11, 2001, and the time is 9:15 A.M., only minutes after attacks on the World Trade Center in New York City and the Pentagon in Washington, D.C. It is an intern's first hour on the job in a federal executive branch foreign policy bureau in Washington. His unit in the bureau, which ordinarily does personnel training, has been preemptively charged with rapidly gathering expert opinion on U.S. response options.

In the weeks and months that follow, the staff (including the intern) set up event simulations and opinion roundtables. Experts are called in to participate in order to develop information needed by the authorities who will decide on U.S. responses to the attacks. The staff's other responsibility is to relay the resulting opinion and suggested options to the policy makers clearly, concisely, and as soon as possible after each roundtable or simulation.

In a series of eight-hour simulations, up to twenty intelligence officers and topic experts project and discuss possible events while the intern observes from a control room. He quickly writes notes as significant occurrences or ideas develop. Other staff members observe the simulation from other control rooms. Immediately afterward, while memory is fresh, the individual observers flesh out their notes. Approximately twenty pages of expanded notes result for each observer (eighty pages total for a simulation).

Also, in another series of day-long roundtables, ten to fifteen academic or government experts and foreign policy professionals make timed presentations followed by intensive discussion while the intern and staff meet in the room with participants to transcribe the talk. Transcriptions combine word-for-word quotations, terse summary, and the individual transcriber's personal code signifying key topics. Immediately after a roundtable, each transcriber deciphers and condenses the original transcript. Reduced to essentials, approximately twenty pages result for each transcriber (eighty pages total for a roundtable).

Following each event, after eight or more hours of intensive observation, note-taking, and transcription, the staff work on into the night. They compare four sets of observation notes on a simulation to identify the most important occurrences and ideas. Or they compare four transcriptions of a roundtable to identify the most important topics and opinions. Even later, the unit supervisor and staff roughly outline a memo, and a staff member drafts it. The next day, the supervisor and staff (including the intern) revise the draft memo until all agree that every word is factual, is pertinent, and cannot be misconstrued. The finished one- to two-page memo is delivered two to three days after the event.

Scenario 2

Local government officials anticipate a farmer's request to operate a concentrated animal feeding operation (CAFO) in the municipality. It will be the first request of its type made since the municipality adopted a comprehensive plan for land use and a zoning ordinance. In that ordinance, although it is ambiguous, CAFOs may be considered as conditional uses of agricultural land to be permitted only if they meet site-specific conditions.

In preparation for reviewing the anticipated request and making a decision, the municipal official who chairs the planning commission begins to self-educate regarding CAFOs. His first objective is to become familiar with state law governing local authority to regulate CAFOs. He browses county and state government listservs; identifies technical and legal experts for possible consultation; attends relevant workshops, public hearings, and meetings; searches databases maintained by the state association of municipal officers and state government agencies; and searches municipal archives of public comment during the process of zoning adoption.

In a public meeting, he offers a preliminary interpretation of state law setting boundaries on municipal authority regarding CAFOs. A resident attending the meeting questions his interpretation. After the meeting, she offers to research the matter further. Given his lack of staff and the limited time he can devote to any single problem, he accepts her offer.

She locates the relevant state law and regulations online and reads them. She telephones state officials involved in authoring and implementing the regulations to ask about interpretations. They refer her to current case law on CAFOs and municipalities. Following up on their referrals and using the help of librarians in the state law school, she reads synopses of relevant current and pending cases.

Several days after the meeting in which she raised questions, she composes a one-page summary of findings and includes her interpretation of them. She e-mails the summary to the local official.

Policy makers need information for making decisions. They usually prefer it in short, quickly comprehended summary form. This chapter helps you to write two forms of summary, a briefing memo and an opinion statement.

During a policy process, authorities receive large amounts of unsolicited information and advice. Often, they ignore it. Instead, policy makers directly seek the information and advice they need.

What kinds of information or advice do policy makers typically need? For consideration of an issue, general information might include assessments of events or conditions; arguments and critical

analyses of arguments; reviews of policy options and technical analyses of the options; specialized topic reports; investigative reports;
summaries of laws germane to the issue; legal counsel on interpretation of laws; and summaries of expert opinion, public opinion, and
political advocacy. Beyond these general types of information, any
single issue demands its own particular and detailed information.

For example, a municipality that is developing a comprehensive
plan for land use will need general assessments of area conditions
(environmental, economic, historical, and cultural factors), reports
on current costs of providing services in the area (such as roads, water, and sewage treatment), summaries of relevant state laws (such as
regulations governing municipal planning), and more. To apply general information to a specific municipality (such as a township or
village), its elected officials might ask county or state government
agencies for local population statistics, economic projections, or environmental data. They might ask legal counsel to examine land-use
planning tools such as zoning ordinances in nearby municipalities or
to review case law on legal challenges to them. To prepare for public
discussion of draft plans and ordinances, the officials will seek political advice. They will want to know the opinion and advice of organized groups as well as individuals living in the municipality.

Who provides information to policy makers? It varies by level of
government. In federal and state governments, professional staff
might produce much of the needed information. The staff's know-
how, or familiarity with the policy process and understanding of
the political context, enables them to inform policy makers usefully.
Staff members typically write briefing memos. As distinguished from
extensive memos such as policy analysis memos (see chapter 3),
briefing memos are terse and targeted summaries of essentials. They
might be composed to update an official on a current topic, for
example.

Because municipal governments often have small staffs (or no
staff), local elected officials might do their own information gathering. They might utilize a range of information providers including
experts (representing subject knowledge), advocacy and stakeholder
groups (representing organized interests), legal counsels (representing rules and procedures), other officials and associations of elected
officials (representing politics), and citizens (representing the opinion or experience of individuals or groups). Any of these providers
might write an opinion statement or a briefing memo to inform an official's work of representation.

How to Inform Policy Makers in a Briefing Memo or Opinion Statement

Goal: Recognition of meaningful information in a mass of details and representation highlighting the significance of information for a user.

Objective: Skills of distilling, listening, recording, observing, evaluating sources, relating details to context, interpreting details accurately in context, and selecting details according to relevance; capability of stating informed opinion that is aware of and responsive to other opinions.

Product: One- to two-page written memo, or one- to two-paragraph written statement, possibly with attachments.

Scope: Only essential topics in an identified context to target a specific information need.

Strategy: Use of guiding questions to develop the memo's or statement's contents:

- Why is this communication necessary? Consider the workload (and information overload) of the user. Is the communication needed at all? Is it needed now?
- What is the subject? Consider the interruptions and distractions in the user's routine. Exactly what is the communication about?
- What is the purpose? Do you want to inform policy makers about their choices? Persuade a community? Help the general public make up their mind? Lobby influential actors in order to influence outcomes? What must the document include and exclude, based on your purpose?
- What will this communication do? What can happen as a result? Consider what you want recipients to think or do. What must the document include or exclude to enable the intended action? What other consequences (other than the intended) might the communication have?
- What is the context? To what policy process does this communication relate? Who are the players? What must the document include or exclude to catch the intended recipient's attention?
- What is the situation of reception? Consider who will read and use the document. How do they like or need information to be presented? Will they refer it or forward it to others? In what circumstances will it be read or used?

How will you design the document for readability and usability?

- Why are you providing the information? Consider the ethics of your communication. Do you have a position on or a stake in the subject? Do you have a role in the policy process? How will you make your interest clear to the recipient?

Task #1. Develop the information.

From meetings:
- Attend relevant public or private meetings; take full notes; get copies of statements and other documents available; get contact information for participants.
- Get a copy of the agenda (if there is one) beforehand or at the start of the meeting.
- Attend to what's said and to the context it represents during the meeting.
- Jot (in the margins of the agenda) your own notes and questions about the proceedings, and capture (as nearly verbatim as you can) the significant questions asked by others.
- Contact participants, government staff, topic experts, or knowledgeable citizens for answers to questions or referrals to sources immediately after the meeting.
- Consult the sources to gain a better sense of the context.

From varied sources:
- Find and read pertinent publications and materials.
- Consult with knowledgeable people inside and outside government:
  - Professionals who draft legislation, administer law, or litigate issues
  - Librarians who are in government information collections
  - Academic specialists or practitioners who are in relevant fields
  - Nonprofit advocacy organizations that are interested in the issues
  - For-profit organizations that are interested in the issues
  - Journalists who inform the public about the issues
  - People who are actually or potentially affected by the problem in everyday life

- Take notes on your conversations; record your search paths; save copies of e-mails.

From informed reflection and analysis:
- Update the original questions and reframe the issues as information develops.
- Pause periodically to summarize your understanding and to critically examine it.
- Continue to consult as needed to improve your understanding of the process and context.

Task #2. Write the memo or statement.

Before you write, review the method in chapter 2. To target your purpose and audience, substitute questions in this chapter (see Strategy, above) for the method's more general questions, as appropriate.

Craft the document's contents for quick comprehension and ready use. Do not include everything you know; include only what the user needs and what the purpose requires. (You can provide more information later, if necessary.)

Choose the right presentation. If you are representing an organization, use its template (if it has one) for memos or statements. Communicate your memo or statement on the organization's letterhead stationary. If you are free to design the communication, fit it into one or two pages (or the equivalent). Provide a header (to/from, subject, and date), an overview sentence or paragraph, and chunks of text with subheadings that summarize the contents of each chunk. You might also use a cover letter or attachments. Note: Before attaching anything crucial, consider the circumstances of reception, or how the document will be read and used. Attachments sometimes get detached when the document is circulated.

After drafting the communication, review and revise as needed (see method and checklists, chapter 2). If you are pressed for time, revise only to focus the message sharply. From the reader's perspective, that is most important.

Two Examples

Example 1.   Here is a London newspaper's summary of investigations into the decision in 2003 by Tony Blair (United Kingdom prime minister) to join George W. Bush (U.S. president) in a war against Iraq. Investigation was prompted by the apparent suicide of a UK government staff member involved in the decision making.

INVESTIGATION SUMMARY

Clarity emerges as court closes its doors: What has been learned on 10 key points as inquiry's first phase ends

The Guardian, Saturday September 6, 2003

The doors of Court 73 closed on Thursday after four weeks in which the inner workings of the British government were laid bare as seldom before. Lord Hutton will return to the Royal Courts of Justice a week on Monday. In the first phase of the inquiry a wealth of emails, memos, letters, minutes, and personal testimony have revealed several important truths about the government's handling of the case for war and its impact on one man, David Kelly.

1. The many faces of Dr. Kelly
The portrait emerging is of a complex, self-contained individual, a man enthusiastic about his work but uncommunicative in private life who found it difficult to share his personal worries. An intimate description of the man was given by his widow Janice, daughter Rachel, and by friends and colleagues.

His religious beliefs—the Bahai faith—and his scientific background committed him to telling the truth. But inconsistencies have been thrown up. He told the Commons foreign affairs committee he had not spoken to the BBC reporter, Gavin Hewitt, and he failed to acknowledge a phone conversation with another BBC reporter, Susan Watts.

2. Dr. Kelly apparently killed himself having come under growing pressure in the glare of public attention
The evidence from his widow, his daughter and Keith Hawton, director of the centre for suicide research at Oxford University, created a vivid picture of a man under growing pressure. Mr. Hawton

concluded that he almost certainly did commit suicide and that, though the causes of suicide are multi-faceted, he had not made the decision until the day of his death.

One factor was the continuing emails and calls from the Ministry of Defence that may have led him to realize that his evidence to committees of MPs did not mark the end of the affair but instead meant he was being drawn further in.

Mr. Hawton said Dr. Kelly had suffered "severe loss of self-esteem" as a result of being described by the MoD as a low-ranking civil servant, and felt a sense of "dismay at being exposed to the media."

3. Downing Street took very seriously the BBC's allegation it had inserted the 45 minute claim into the Iraq dossier knowing it to be wrong
Lord Hutton has heard that for Tony Blair the allegations could not have been more serious: "You already have this extraordinarily serious allegation which, if it were true, would mean we had behaved in the most disgraceful way and I would have to resign as prime minister," Mr. Blair told the inquiry.

It amounted to lying to parliament, and Mr. Blair said he viewed the BBC report as having accused him of having duped the public into the war. The PM admitted the festering row had derailed his agenda: "Since then that has been the issue. I mean we are three months on and it is still the issue."

4. No. 10 was keen to see Dr. Kelly's name become public partly to forward its case against the BBC
Alarm bells rang in Downing Street and the MoD when Dr. Kelly admitted speaking to the BBC's Andrew Gilligan. As soon as they suspected he was Gilligan's source, they wanted to use him in their battle with the BBC, in particular to say that Gilligan had embellished what the scientist had told him.

The MoD gave clues to journalists about his identity before confirming his name. Dr. Kelly, a very private man, was told to give televised evidence to the Commons foreign affairs committee, an experience from which he never recovered, the inquiry heard. Lord Hutton has made clear that the outing of Dr. Kelly, and the way the MoD treated him, will feature strongly in his report.

5. Tony Blair was intimately involved in the naming strategy of Dr. Kelly
Mr. Blair was told on July 3, by Jonathan Powell, his chief of staff, that a possible source for Gilligan's story had emerged.

The prime minister's weekend at Chequers was peppered with further discussion about the matter with his most senior officials, including Alastair Campbell. On Monday July 7 Mr. Blair met officials in his study, and asked what was known of Dr. Kelly's views and what would he say to MPs. A decision was made to have him reinterviewed.

On Tuesday the Kelly issue arose at three meetings involving the PM. The decision to issue a press release was approved and written by Mr. Blair's spokesman and other top officials.

On July 9 the PM's official spokesman revealed a crucial detail about the identity of the supposed source which helped to lead reporters to Dr. Kelly.

6. The Iraq dossier on weapons of mass destruction was over-egged
Compare a copy of the draft dossier on September 5 and the final publication on September 24 and there is no doubt that it was hardened up.

The draft, which had existed since February last year, was a typically dry Foreign Office/MI6 document, with the emphasis on how Iraq had these weapons before 1998. The final document uses the same information but the emphasis has switched to such weapons presenting a current threat.

Brian Jones, a former analyst with the MoD's defence intelligence staff, claimed it had been over-egged. Mr. Campbell denies sexing it up, but the changes he suggested did alter the presentation by portraying Iraq as a bigger threat than it appeared to be on September 5.

7. The claim in the dossier that Iraq could deploy its chemical weapons within 45 minutes of an order to do so came from a second-hand and dubious source
The claim was controversial before the inquiry, but now looks very tarnished. The government said it came from an Iraqi officer. Possibly, but this is not quite the whole truth. In fact it came second-hand through another source.

Staff in defence intelligence, according to one of its former top analysts, feared the motive of the source who reported the claim, worrying he might be trying to influence rather than inform. Even intelligence officials who thought the dossier a good idea were worried about the 45 minute claim. Dr. Kelly thought it "risible" that WMD could be deployed so soon, one witness said.

John Scarlett, the chairman of the joint intelligence committee, let slip [that] the claim referred to battlefield munitions, and not missiles with a longer range. The dossier led to claims that Iraq could launch WMD within 45 minutes towards British bases in Cyprus.

8. There were rows within the intelligence community over the wording of the dossier
Intelligence officials, notably experts in the MoD, questioned many of [the] Iraq weapons dossier's key assertions. Dr. Kelly shared these concerns, but they were ignored by those drawing up the dossier under the overall supervision of Mr. Scarlett, who developed a very close relationship with Mr. Campbell.

Mr. Scarlett insisted that he, and not Mr. Campbell, retained "ownership" of the dossier until its contents were agreed. But the inquiry heard that the full JIC did not approve the dossier, which was drawn up by a special group including Mr. Campbell. The full extent of concerns about the dossier were not aired in the inquiry since MI6 officers—who provided much of the raw intelligence—did not give evidence.

9. Geoff Hoon was less than fully truthful in his evidence to the inquiry
The defence secretary's evidence to the inquiry saw him try to deny any responsibility for how Dr. Kelly, effectively his employee, was treated.

The last revelation of stage one of the inquiry was that he appeared to have been less than candid with Lord Hutton. Mr. Hoon had been at a key meeting to approve the naming of Dr. Kelly, his special adviser, Richard Taylor, revealed.

The defence secretary had failed to mention this in his evidence, during which he said he had not been involved in any talks about confirming Dr. Kelly's name to the media. An admission from Mr. Hoon that he knew about the decision to confirm the scientist's name had to be dragged out of him by the inquiry counsel James Dingemans QC.

10. Andrew Gilligan's story that No. 10 wanted to "sex up" the Iraq dossier has largely been vindicated, but his claim that Downing Street

inserted the 45 minute claim knowing it to be wrong has not been substantiated
The claim by the BBC journalist, based on a meeting with Dr. Kelly, that No. 10 wanted to "sex up" the dossier has been largely supported by evidence.

The inquiry has heard that defence intelligence officials believed the dossier was "over-egged" after pressure from outside, taken to mean No. 10. But there is no evidence to back Gilligan's initial claim that the government knew the 45 minute claim was wrong.

The inquiry also heard that Gilligan wrote an email to David Chidgey, a Liberal Democrat member of the Commons foreign affairs committee, outing Dr. Kelly as the source of a Newsnight story by his BBC colleague Susan Watts.

Gilligan was also accused at the inquiry of playing a "name game" with Dr. Kelly, trying to get the scientist to identify Mr. Campbell as responsible for "sexing up" the dossier.

"Hutton inquiry: Clarity emerges as court closes its doors: What has been learned on 10 key points as inquiry's first phase ends," Richard Norton-Taylor, Vi Kram Dodd, Ewen MacAskill, 6 September 2003. Copyright Guardian Newspapers Limited 2003.

What This Example Shows.   This journalistic summary is crafted for a general public audience. It might also serve to brief elected officials on its topic, the detailed investigation of a complex government decision. (This newspaper article might be circulated in UK and U.S. government offices to brief anyone who needed to know the results of this investigation. Major newspaper accounts are sometimes used for briefing purposes if they succinctly provide necessary information. See method, chapter 2.)

While its current length exceeds the one to two pages preferred for brief memos, the example could easily be condensed. The top-down pyramid organization of contents and the subheadings that serve as mini-summaries enable compression of the article's information to a list of "talking points" for an oral briefing, for example. Clarity in the original is enhanced by the subheadings and the concise wording (British style). Altogether, this article's effective document design includes:

- A headline previewing the contents
- A pyramid layout presenting information on the main topic first
- An organized list containing the most significant conclusions

- Each conclusion being stated in a subheading
- Subheadings being presented in a declarative "who did what" sentence
- Each conclusion containing key supporting details concisely summarized immediately following the subheading

Example 2. Here is a memo written by the community resident to the local official described in Scenario 2 at the beginning of this chapter. The memo was sent as an attachment to an e-mail message.

E-MAIL MESSAGE

Doug:

I recognize that the township supervisors and planning commissioners are trying to operate within state law on CAFOs [concentrated animal feeding operations]. It's a complicated task, and I appreciate the careful thought you are putting into it. My intentions are to help by getting good advice on interpreting state regulations so that we rightly know what authority Gregg Township has regarding CAFOs.

You raised two concerns in the July meeting of township supervisors. In the attached memo, I report what I've found and what I think regarding those two concerns. Basically, I find and I believe that the township has the necessary standing and authority to regulate CAFOs. You might want to talk to Douglas Goodlander (my source in Harrisburg). I'll call you later this week to see if you want to talk about any of this.

Catherine
ATTACHMENT

Memorandum

Summary
Gregg Township has 'zoned in' the possibility of CAFOs, and it may regulate them within the limits of state law.

Prohibition of CAFOs
You say "You can't zone concentrated animal feeding operations out [prohibit them by means of zoning]." *Reference:* Pennsylvania State

Association of Township Supervisors. See July issue of *PA Township News* for article "Avoiding Controversy: How Townships Can Minimize Conflicts Between Residents and Intensive Ag Operations."

My view: Gregg has provided for the possibility of CAFOs operating in the township by making 'feedlots' a conditional use in the agriculture zone. If 'feedlots' can be construed to include CAFOs, then Gregg has 'zoned in' the possibility. The zoning ordinance has a procedure for permitting, or not permitting, conditional uses based on case-by-case review for specified criteria. *Source:* Gregg Township Zoning Ordinance, Article 3, Agricultural Zone, Conditional Uses, C.4 Feedlots (p. 10), and Article 9, Conditional Uses, C. Criteria (pp. 3–4).

Municipal Regulation of CAFOs

You say "I read the state regulation as taking the wind out of our sails in regulating CAFOs." *Reference:* The "pre-emption of local ordinances" provision of the Rules and Regulations, 25 PA Chapter 83 Subchapter D, 803.25 (b) for the Nutrient Management Act (3 P.S.1701 et seq.) published in the *Pennsylvania Bulletin,* Vol. 27, No. 26, June 28, 1997.

My view: municipalities can regulate CAFOs. The next section of 83.205 (part c) states that "nothing in this act or this subchapter prevents a municipality from adopting and enforcing ordinances or regulations that are consistent with/no more stringent than the state act." Beyond nutrient management [anti-pollution measures to prevent excess nutrients in animal manure from entering water sources], other issues presented by CAFOs such as odor, noise, air pollution, and road use can be locally regulated. *Source:* Douglas Goodlander of the Pennsylvania Conservation Commission. Goodlander is an author of 25 PA [nutrient management regulations] who has been involved in CAFO court cases in PA. He cautions that he can provide interpretation but not legal counsel. He encourages you to call him if you want (telephone number provided).

What This Example Shows.   Two characteristics of this communication are especially noteworthy: its form and its intelligibility in context. Electronic communications such as e-mail differ in several ways from written or spoken communications. Particularly, e-mail is typically more informal and personal and often has less regard for conventions for grammatical correctness. In a public process, if taken out of its original context, e-mail might embarrass the sender or recipient. Or it might compromise the action.

The intelligibility of this message relies on context. The wider context includes both ongoing discussion between the sender and the recipient about the topic and familiarity as members of a shared local community. The sender and receiver know each other and have worked together before. They generally agree on the importance of local control in governance, but they might disagree on regulation of CAFOs. In the limited context of a policy process underway, they have been meeting and talking about this particular issue prior to this e-mail. When the e-mail and its attachment are read together, they can be accurately understood in context. If they are read out of the limited policy context or if the two documents are separated, the communication might be misinterpreted.

Why does this matter? Correspondence with an elected official is open to public scrutiny. In addition, the e-mail and its attached memo will become part of a permanent public record. The elected official is obliged to archive all communications on public matters and to make them available to anyone with a right to public information now or in the future. Even if purged from archives, e-mails might be recoverable, as investigations of government action remind us (see method, chapter 2).

Context is influential here in another way, too. If the two parts of the communication become separated in a process of referring or forwarding the documents, the tone of either one might lead to misinterpretation. The brisk tone and terse wording of the memo might especially seem unfriendly. (The method in chapter 2 advises you to always consider the attitude a communication might convey.) Contents are concisely organized to highlight contrasting interpretations ("You say, I say"). And contents are concisely worded to include only essential findings. Citations and contact information are strategically included to save time for a busy official who does much of his own research in case he wants to consult the references himself. There is a downside to this efficiency, however. The attachment's terseness lacks the sympathetic tone of the covering e-mail that acknowledges the difficulties the official must resolve. If the two documents are received separately, this might make a difference in their reception.

Communication is not only an information exchange; it is also a social interaction. The tone of a policy communication can determine its reception. Recalling that policy makers receive lots of unsolicited communications, you should be aware that tone explains why some communications are ignored. Memos or statements that are expressly hostile or that seem closed-minded are less likely to be read, to receive a response, or to be useful to the process.

More Examples. Here are several more examples that illustrate good tone and bad tone in statements of opinion. These examples were e-mailed to an elected official in county government regarding a proposed merger of city and county schools.

Good Tone. The following three opinion statements got the policy maker's careful attention and received a substantive reply.

A. You currently face a difficult decision regarding the proposed merger of Chapel Hill/Carrboro and Orange County Schools. I am writing to suggest a public referendum on this matter given the significant impact that the results of your decision will have on your constituency. Thanks for taking time to consider this request.

B. The merger discussion is heating up quickly, and I'm hoping the real issue of the disparate funding for the two systems doesn't get lost in the commotion. The push for a referendum, called for by so many Chapel Hill parents, seems a veiled attempt to simply stifle discussion, allowing the real issue to again get swept under the rug, still unfixed.

C. Here are a few questions I'd love to have answered. I know you're busy and probably receiving hundreds (??!!) of e-mails daily on this issue. I hope you can fit me in.
   1. Do you see a funding imbalance between the two systems?
   2.–7. [Subsequent detailed questions not shown here.]
   We need a solution. Thank you for considering my questions.

Bad Tone. In contrast, the following three opinion statements (like many similar ones generated by a letter-writing campaign) got little attention and received no reply.

A. I understand that the Orange County Board of Commissioners is evaluating a merger of the Chapel Hill and Orange County School Systems. I would like to communicate that:
1. I, along with most of my local colleagues and neighbors, are vehemently OPPOSED to a merger.
2. I request that a public REFERENDUM be held on this issue ASAP.
3. Unless proper procedure is followed throughout, a proposed merger will be challenged in the NC and Federal courts to the extent necessary.
4. The voting records of the entire board will be well remembered and publicized in time for the next ELECTION.
B. I am greatly disappointed in your decision Wednesday night to short circuit democracy in our county. None of you ran last November with a position on school merger. You have suddenly sprung it on the citizens of the county. Since you would not face voters on the issue, you should allow a referendum on the issue in the county. Otherwise, you should delay the issue until an election year, and run on your beliefs. The idea that you can have a 'stealth' merger of school systems and avoid the will of the citizens of the county, as some of you seem to believe, is not in keeping with the traditions of transparency and progressive politics in our county. I voted for you all last November. But I did not vote for school merger. Now I feel that your election was as much a sham. I would like a chance to vote on school merger or to vote again on your positions on the county commission.
C. I am very concerned about an article in the *Herald* which indicates that the schools in Orange County and Chapel Hill may merge. I don't understand what the advantage of such a move would be.
   If there is an advantage to the move please let me know what it is.
   If there is no advantage to the move, please let me know by ignoring this message.

What All These Examples Show. A lesson to learn from all these examples is that a public policy communicator has to make many competing choices. Purpose, contents, presentation style and tone, medium of delivery, and concern for immediate reception and use as well as the permanent record—all these must be considered and choices must be made each time you communicate (see method, chapter 2).

All the choices are important because the consequences can be so significant. Communication affects the process and the outcome of public policy activity. The process, the related communications, and the outcome affect people and places in real ways.

❖ 8 ❖

TESTIMONY:
WITNESS IN A PUBLIC HEARING

Scenario 1

A student advocate of nutritional labeling on fast food agrees to testify in a (simulated) federal government hearing on current health issues. He reviews his legislative history research, position paper, and policy proposal on the topic to select a focus for testimony that is suitable for the purpose of this hearing. He will testify before a mock Senate committee with broad jurisdiction for policy on health, education, labor, and social insurance. He plans testimony to represent a position that (he knows from his research) is taken by a nonprofit advocacy organization for consumer education. The position emphasizes failure to enforce judicial decisions requiring food labeling. Enacting the role of spokesperson for that organization, he testifies for two minutes to advocate extending current law to include the fast-food industry. During the question-and-answer period that follows, he responds to both friendly and hostile questioning.

Scenario 2

A resident of a rural area known for the high quality of its cold-water fishing streams serves as a board member of a local conservation group. She learns of proposed changes in state policy for classifying waterways according to their quality. Four regional field hearings will be held to hear public comment on the agency's draft of a revised guidance manual for implementation of state water quality regulations.

With the help of the state agency's staff, the resident obtains copies of the old and the new guidance manual, and she carefully reads the proposed changes. In her judgment, the new guidance draft weakens the standards. She alerts the local conservation group for which she is a board member. She asks the board to authorize her as spokesperson for the group in the upcoming hearings, and they do so. She writes testimony stating the position. She telephones the agency staff and asks to testify. In the call, she provides her credentials and says whom she represents. The staff member who is organizing the field hearings puts her on the schedule of witnesses.

In the hearing, she testifies with other witnesses from throughout her region. All witnesses provide a written copy of their testimony for the public record of the hearing. Following the testimonies, witnesses are encouraged to question the agency's managers attending the meeting. This reverses the usual question-and-answer procedure. Normally, conveners of a hearing question the witnesses. On this occasion, however, the agency managers want to demonstrate more-than-usual responsiveness to public comment.

> *Policy makers and administrators are required to deliberate publicly and to seek public input. This chapter prepares you to testify orally in a governmental public hearing.*

In the U.S. federal government, "sunshine" or public access laws mandate open hearings for all legislative functions—making law, appropriating funds, overseeing government operations, investigating abuse or wrongdoing, and approving nominations or appointments to office. Hearings are held in executive and legislative branches of federal government. In state and local governments, public deliberation is mandated, but formal hearings are not as commonly held as at the federal level.

In government organized by a political party structure, the majority party (the party in power) chairs committees and thus sets the agenda for committee work, including public hearings. Committee chairs (with their staffs) decide whether to hold a hearing on a topic within their jurisdiction, what the purpose of a hearing will be, and who will be on the witness list. Topics and purposes of hearings reflect the committee's policy jurisdiction and the (majority party) chair's political agenda. The agenda might or might not reflect cooperation between the majority and minority interests.

Several committees might hold hearings on different aspects of the same topic, especially if the topic concerns a hot issue that crosses jurisdictions. Hot issues are those that are currently in the news, controversial, or especially significant in some way. Most hearings are not about hot issues, however. Most hearings are workaday sessions to oversee government operations, to decide on appropriations of funds, to reauthorize programs, and so forth. They do the routine work of governance.

Public affairs television usually does not broadcast these routine hearings. Selected daily hearings are summarized on the government page of newspapers and some advocacy group websites.

In the executive branch, departments or agencies hold public hearings on issues within their regulatory responsibility. Some are held in the field, in geographic areas or political districts directly affected. Executive branch hearings vary in format from informal public meetings to formal deliberative sessions. The state environmental agency's hearing on water protection described in Scenario 2 (above) illustrates informal field hearings in which executives, staff, and witnesses freely discuss a topic.

Taking U.S. congressional committee hearings as the model, hearings typically follow this order of events. After the chair opens the hearing, announces the purpose, and states his or her position on the topic, the committee members then state their positions and, possibly, their constituency's concerns regarding the topic. Next, invited witnesses testify on the topic. Following the testimonies, the usual practice is for committee members to question the witnesses.

In principle, anyone might be invited to testify who can provide information that lawmakers or administrators seek. In reality, witnesses testify at the federal level only by request of the committee. At state and local levels, the witness list is more open. There, you may be invited, or you may ask, to testify. If you wish to testify, you contact the staff of the committee or the agency holding the hearing.

In the communication situation of a typical hearing, witnesses testify as spokespersons for an organization or a government agency.

Occasionally, individual citizens testify on their own or their community's behalf. Witnesses must relate their special concerns to a policy context and their agendas to other agendas. Policy makers and witnesses interact face-to-face, and exchanges might be polite or confrontational. Questioning might be focused or loose. Questioning is always political, and sometimes it is bluntly partisan. The atmosphere might be orderly or hectic. The time limits are always tight— typically one to five minutes for each witness to present testimony and five minutes for each member to question all the witnesses. There might be multiple rounds of questioning. Hearings can last for hours or days if the committee or the witness list is large.

Everything communicated in a hearing goes (via a legislative stenographer) into a transcript. This transcript is the official public record of the hearing. There are actually two public records, unofficial and official. Unofficially, the hearing might be broadcast and reported by news media. These are influential accounts that significantly shape public discourse and the perception of problems; however, they are not authoritative. They would not be included in a legislative history, for example. For the authoritative and official record of a hearing, a stenographer records the statements, questions, and answers verbatim, exactly as they are given. In accord with current legislative reporting practice, the verbatim transcript cannot be edited, except to correct factual errors. The transcript is later (sometimes months later) printed and published by the superintendent of government documents through the government printing office. This is the official, or legal, record of the hearing. Published hearing records are important for democratic self-governance because they give continuing public access over time to the accurate and full information produced by a hearing. That information is useful for many purposes. Journalists, law clerks, academic researchers in many fields, legislative staff, lobbyists, advocates, and active citizens use hearing transcripts as sources. Published hearings are primary sources for legislative history research, for example. They are also major sources for determining a law's original intent when the law is being adjudicated.

In the overall significance of government hearings for democracy, witness testimony is perhaps most important. For witnesses, it is an opportunity to make personal or professional knowledge useful for solving problems. For policy makers, it offers a rare opportunity to interact directly with knowledgeable people and to question them. Policy makers appreciate that interaction. Most information they receive is filtered through staff or advisors. They like having the chance to interact directly with information providers.

How to Deliver Oral Testimony Based on a Written Statement

Goal: To speak authoritatively and to answer questions responsively in public deliberation.

Objective: Skill of writing speakable text, skill of speaking easily from written text, and readiness to answer anticipated and unanticipated questions.

Scope: Pinpointed topic pertinent to a hearing's purpose and the witness's role.

Product: Two expected communication products:
- short oral summary, either a list of talking points (outline for speaking) or a one-page overview (to be read aloud)
- full written statement, possibly with appendices, to be included in the record of the hearing

Strategy: Confident and useful public testimony resulting from advance preparation.

Developing a Testimony.   Obviously, witnesses must know their subject and their message. More importantly, witnesses must understand the purpose(s) of the hearing and their own role and purpose(s) for testifying. Effective witnessing is achieved by presenting concisely and by responding credibly to questions. Responding to questions effectively is most important. If you are on the witness list, you are acknowledged as having something relevant to say. You do not need to impress people by showing how much you know about the topic. Focus strongly on your purpose and your message in relation to the hearing's purpose.

Know the Context.   To what policy process does the hearing relate? To what political agenda? Who's holding the hearing? What is the stated purpose of the hearing? What is the political purpose? Who else is on the witness list? What are their messages likely to be?

Know Your Message.   Distill your message into one to two sentences that you can remember and can say easily. How does your message relate to purpose of the hearing? How does it relate to other witnesses' messages? Anticipate committee members' responses and questions. What are you likely to be asked?

Know Your Role. Are you speaking for an organization? For yourself? Why are you testifying? What do the organizers of the hearing hope your testimony will accomplish?

Know the Communication Situation. Will the press attend the hearing? Are you available for interviews after the hearing? Will the hearing be televised? How is the hearing room arranged? Do the arrangements allow you to use the charts, posters, or slides? Are those visual aids a good idea if the room lights cannot be dimmed (due to televising the hearing)? What is the location for the hearing? If you are using charts, posters, or slides, how will you transport them? Who will set them up in the hearing room?

Rehearse Your Delivery. Will you read your statement or say it? Generally, saying it is preferred. Be ready to do either, however. Rehearse by reading the full statement aloud and by speaking from an outline. You'll discover which way is easier for you and which you need to practice more.

Task #1. Write the testimony.

Use the method in chapter 2 to plan testimony in both oral and written forms. Some witnesses prefer to outline the oral summary first and then to develop the full written statement from that outline. Others prefer the opposite way. They write the full statement first; then they outline an oral summary based on the written statement.

The key is to prepare both. Write out your oral summary, even if it is simply a list of talking points on an index card. The written list will provide confidence and control as you testify. Recall that everything said in a governmental public hearing is recorded and that the record is made publicly available. Do not plan to wing it or to testify extemporaneously. If you do that, you risk exceeding time limits, which committee chairs do not like. And you might forget important information, say more or less than you intend to have on the public record, or find yourself being asked questions (about something you said) that you are not prepared to answer.

If you are free to organize your testimony, use this sample template. Use it in outline form for the oral summary, and expand it appropriately for the full written statement. Put extensive support in appendices, not in the main statement. (Both the oral and written versions will be included in the transcript of the hearing.) Here is the template:

- Title page or header to identify the organization and the witness, the agency holding the hearing, the topic, the date, and the location of the hearing
- Greeting to thank the organizers for the opportunity to testify and to state why the topic is important to the witness
- Message to state the main information the testimony provides
- Support (evidence, grounds) for the message
- Relevance of the message to the hearing's purpose
- Optional: discussion or background to add perspective on the message (only if relevant or if specifically requested by conveners of the hearing)
- Closing to conclude the testimony and invite questions

Task #2. Write the full statement.

The written statement might use the same organization as the oral summary. The written statement may be longer, include more details, and be accompanied by appendices. It can be any length, but it should be no longer than necessary. Even if the written statement will be lengthy, it must be concisely organized and worded. That way, it is more likely to be used.

Task #3. Present the testimony.

The following tips are important.
- *Summarize.* During oral delivery, whether reading a document aloud or speaking from an outline, state only the essentials. Save the details for the question-and-answer period.
- *State the message early and emphatically.* Whether reading a text or speaking from an outline, state the message up front.
- *Stay within time limits.* Usually, the chair of a hearing will tell you the time limits. If not, assume that you have two to five minutes for a summary. Do not go over the limit.

• *Listen*. Closely attend to the opening statements by the committee chair and the committee members. Opening statements cue the questions that you might be asked. Or they might include content to which you want to respond later, when it is your turn to speak. Listen also to other testimonies. Committee members might ask you to comment on other witnesses' testimonies.

• *Answer credibly*. The question-and-answer time is often the most important part of a hearing. Committee members and witnesses alike agree on this. For committee members, it is a chance to interact directly with knowledgeable people. Members usually ask prepared questions in order to get important concerns, as well as witnesses' responses to the concerns, on the record. For witnesses, the question-and-answer time is a chance to connect their message to varied agendas represented in the questions or to pinpoint the usefulness of their knowledge to the committee. Witness effectiveness depends primarily on the witness's credible (honest, accurately informed, relevant) responsiveness to questioning.

After you have presented your testimony statement, shift your attention to question-and-answer communication. Follow these important guidelines:

• Listen to the questions asked of other witnesses. Do not daydream or otherwise lose focus while others are being questioned.

• Make sure you hear each question correctly when you are being questioned. If you are not sure you heard, the question correctly, ask to have it repeated.

• Answer the question that is asked, not some other question that you half-heard or that you prefer.

• Stop when you have answered a question. Wait for a follow-up question. Postpone details, elaboration, or qualification on your original answer until a follow-up question allows you to provide them.

• Do not lie or invent information. If you hear yourself fabricating an answer (perhaps out of nervousness), stop. Politely ask to have your answer removed from the record, and begin again.

• Handle these situations especially carefully:
  − *You are asked for your personal opinion*. When you testify as spokesperson for an organization, be careful to

present the organization's viewpoint. Avoid giving a personal opinion unless specifically requested, and then only if you appropriately can do so. If you do, be careful to distinguish your own view from the organization's. Alternatively, politely restate the organization's message, and say that you are more prepared to discuss that.

- *You don't know the answer.* Depending on the dynamics at the moment (neutral or friendly or confrontational) and considering the effect on your credibility of not answering, you might choose among these options: Simply say you do not know, say you are not prepared to answer but can provide the answer later, ask if you might restate the question in a different way that you can better answer, or defer to another witness who can better answer the question.

- *Your credentials are challenged, or your credibility is attacked.* Do not get angry. Do not confront the challenger or attacker. Politely state your or your organization's qualifications to speak on the topic of the hearing. Restate why you are testifying or why the hearing topic is important to you or your organization. Maintain your role in the hearing as a source of information and perspective not offered by others. Maintain your composure.

Two Examples

Many excellent samples of written testimony can be found on the websites of nonprofit advocacy organizations, public policy institutes, and some government agencies. Here are two sites (a federal government agency's and a nonprofit advocacy group's) that have many testimony samples:

- U.S. Government Accountability Office Reports and Testimonies at http://www.gao.gov/audit.htm
- GRACE Factory Farm Project Speeches and Presentations, Testimonials, and Editorials at http://www.factoryfarm.org/reports.html#gffpspeeches and http://www.factoryfarm.org/takingaction-testimonials.html

Example 1. This example is written and spoken by a student in a classroom-simulated congressional hearing. The student, whose samples on nutritional labeling of fast food have been presented in earlier chapters, is here enacting the role of spokesperson for a policy analysis think tank. In that role, he presents his own message on the importance of labeling. Shown here is the one-page oral summary that he also submitted (with appendices) as the written statement for the record. Not shown here are accompanying charts detailing the impact of nutritional labeling on consumer choice and the compliance with labeling requirements by fast food industries.

TESTIMONY ON NUTRITIONAL LABELING OF FAST FOOD

Senate Committee on Health, Education, Labor, and Pensions (simulation)

Good afternoon Mr. Chairman and Committee members. My name is Nicholas Alexander and I am a spokesman for the [named] Institute. The question that we put before you is whether or not to require major fast food chains to print nutritional information on their products' packaging. We argue that these labels will educate consumers and impact their menu choices to promote healthier eating habits. Opponents argue that nutritional labels do not lead to better diets and that the cost of altering packaging is not justified. We disagree, based on evidence (attached to our testimony) that indicates positive health outcomes of food labeling.

My testimony today will focus on two key issues:

1. The effectiveness of nutritional labeling and mandating labels through federal policy. Studies conducted over the six years following the implementation of the Nutrition Labeling and Education Act of 1989 have shown significant positive effects on national diet and suggest that similar effects may result from imposing labeling requirements on fast food products.

2. The failure of fast food corporations to abide by judicial decisions requiring distribution of nutritional information flyers and posters at all fast food restaurants is concerning. Given the observed effects of these labels on consumers' food choices and

the potential health benefits, this corporate negligence cannot be overlooked.

Thank you. I welcome your questions regarding our testimony. [Accompanying charts not shown.]

Example 2. This example is written by the citizen testifying in state environmental protection agency field hearings as described in Scenario 2 at the beginning of this chapter. Shown here is the written statement for the record (edited to reduce length). An oral summary was delivered.

LIVING WITH EXCEPTIONAL VALUE

Testimony by the Penns Valley Conservation Association (PVCA)
Http://www.pennsvalley.net

Public Hearing on Anti-Degradation Implementation
Pennsylvania Department of Environmental Protection (PA-DEP)
Bureau of Water Quality and Wastewater Management
Harrisburg, PA, August 1, 2001

Thank you for providing us the opportunity to comment on DEP's draft Guidance for water quality protection. We strongly support DEP's anti-degradation program. Because we support it, we are concerned about how local communities such as ours perceive and participate in the program. We focus our testimony on the need to make implementation more inclusive and to ensure public participation. We offer related suggestions for revising the draft Guidance.

Our message to DEP is this: The goal of regulation is water quality protection. To the extent possible under Pennsylvania law, DEP's guidance should assume that implementation requires equal participation by petitioners or applicants and by communities that must live with the consequences of permitted or approved activity. The Department's function is to arbitrate between these parties and their interests while protecting the larger public interest. Petitioners and applicants are well-prepared to present their legitimate interests and the commercial value of granting their request. Communities are less prepared to protect their interests. To carry out its function, DEP must ensure effective public participation.

Addressing those concerns one at a time, and relating them to the draft Guidance:

• *The goal of the program is water quality protection.* The draft Guidance does not make sufficiently clear that the purpose of the anti-degradation program is to *protect* all surface waters from adverse impacts on fish species, flora, and fauna by activities receiving a DEP permit or approval. True, policy is stated in chapter 1 and regulations as well as standards are identified at the start of chapter 2. But discussion sections throughout the draft create doubt that DEP will protect Pennsylvania's resources as required by regulation and federal law [detail omitted].

• *Guidance should assume that protection requires full participation of affected communities in addition to applicants or petitioners. DEP's function is to arbitrate between those interests and to protect the larger public interest.* The draft Guidance focuses exclusively on DEP's response to applicants or petitioners for permits or approvals. Community representatives such as citizens' groups must also be recognized as key participants in permitting or approval processes. The public will refer to the Guidance for policies, definitions, and procedures. The Guidance might function as the procedures manual for public participation, but the current draft does not serve that function well [detail omitted].

• *To carry out its function, DEP needs effective public participation.* According to the Guidance, applicants or petitioners are encouraged to go beyond public notification to seek public input. That is not enough. The Department itself should actively seek and inform community input. On that topic, we must caution DEP about an effort, noted in chapter 4, on the processing of petitions, evaluations, and assessments to change a designated use. We are concerned about the pilot program of notifying landowners who border streams or stream segments being considered for HQ or EV status. That notification is dangerously insufficient. To notify landowners alone—and not local conservation groups, watershed associations, or municipal planning commissions—favors one constituency, property owners. Worse, to notify landowners without spelling out which activities or permits might affect a protected stream is likely to generate misinformed reaction. Backlash against protection is fueled by selective and cryptic public notification.

To summarize, experience teaches our organization that classifying a stream as High Quality or Exceptional Value is relatively easy. But implementation of protection on-the-ground in the community can be hard. We've identified three main obstacles: public ignorance or misunderstanding of the anti-degradation program's purpose and methods, burdensome permitting, and weak coordination among DEP bureaus sharing responsibility for water quality protection.

If time allows, I will briefly describe our experience in attempting to protect special waters and a watershed. Direct practical experience is the context for our testimony [detail omitted].

Conclusion

PVCA applauds the new ground the DEP is exploring in financing watershed assessment and restoration activities by local communities through its Growing Greener program. Conservation groups like the PVCA are adopting watershed-wide approaches to restoration, while monitoring local activities for adverse effects on water sheds. We would like to see watershed concepts reflected in DEP's regulatory and permitting process. We would like support for comprehensive restoration efforts. And, we would like to see stronger inter governmental coordination in protecting special waters.

This concludes our testimony. We refer you to our accompanying chapter-by-chapter list of suggested revisions to the draft Guidance. We are glad to answer questions now.

What These Examples Show.   Example 1 shows good presentation in the role of spokesperson for an organization. The witness's role as spokesperson is made clear. The testimony's contents are condensed, well structured, and concisely worded, with time limits in mind. However, many sentences are too long for smooth oral delivery and for easy listening comprehension. Some sentences should be shorter to increase the speaker's comfort. The message should be emphasized more (see method and checklists, chapter 2).

Example 2 is well organized to support reading aloud under variable time limits. (Sometimes a witness is given more or less time than expected for presentation.) The introduction and summary can be presented in one minute. The introduction, summary, and list of concerns (without details beyond the first sentence in each section) can be presented in two to three minutes. The whole can be presented in five to seven minutes (see Task #1, this chapter). However, as with Example 1, individual sentences are too long for easy speaking and listening. Tighter sentence structure could make sentences here easier to speak and to listen to.

The genre, oral witness testimony (based on a written statement for the record), is a staple of public legislative hearings in democratic governance. It is characteristically more free-form than legal hearings in administration of justice. In government hearings, for example, there are no prescribed rules for disclosing evidence or for objecting to questions as there are in law court hearings. Consequently, witnesses for legislative hearings must prepare well for anticipated as well as unanticipated developments (see method, chapter 2).

❖ 9 ❖

WRITTEN PUBLIC COMMENT: INFLUENCE ADMINISTRATION

Scenario I

A national transportation safety investigative board holds four days of hearings on air bag safety. The board, which reports both to the Congress and to the executive branch, is concerned about the unanticipated high rates of injury from air bag deployment. The witness list for the hearings includes representatives of auto manufacturers, insurance companies, safety institutes, auto safety advocacy groups, air bag manufacturers, and auto parts suppliers. The purpose of the hearings is to enable the board to make recommendations for improving air bag safety.

Based on the board's ensuing recommendations as well as its own investigations, the National Highway Traffic Safety Administration (the federal agency responsible for automotive safety regulations) announces that it intends to modify the current standard for air bags. The agency announces its proposed modification in the *Federal Register* and calls for public comment.

In response, two experts in automotive safety jointly write a technical comment. They point out shortcomings in the agency's proposed modification, and they propose an alternative.

Later, in the preamble to their next published notice on air bag safety, the agency summarizes all comments received. It says how the comments influenced their plans to modify the standard. Here is an extract from that preamble: "In response to the public comments on our 1998 proposal and to other new information obtained since issuing the proposal, we are issuing a supplemental proposal that updates and refines the amendments under consideration." In an appendix, the agency stated its reasons for rejecting the experts' proposed alternative.

Scenario 2

A state environmental protection department's bureau of mining, which regulates mineral extraction industries in the state, announces a proposed revision in a mining company's operating permit. In accord with "sunshine" requirements for permitting processes, the agency publishes the applicant's proposal to mine deeper than its original permit allowed.

The mining company is asking the bureau to remove restrictions on the company's operation at a specific site. The primary restriction prohibits mining at levels that might adversely affect local groundwater quantity and quality. The restriction is warranted in a region where well-water supply varies according to groundwater conditions and where high-quality cold-water trout fishing streams are fed by local springs near the mining site.

In response to the bureau's call for public comment on the mine operator's request for permit revision, a local environmental conservation group, a local civic organization, a local affiliate of a national trout fishing advocacy group, local businesses, and numerous concerned citizens write letters. At the request of the civic organization, the agency holds a public meeting. The conservation group hires a professional stenographer (who is also a notary public) to transcribe the meeting. In addition, the group invites local news reporters. The meeting is well attended. The bureau officials, the mine operator, and the residents of the region affected by the mine vigorously discuss the request to lift restrictions on the mine. After the meeting, the conservation group provides the transcript to the bureau as a written record of public comment. If there is litigation regarding this permit, the transcript will provide evidence.

In announcing its decision later, the agency says, "The many public comments the Department received regarding this application formed the basis for modifications to the permit revision and resulted in changes in [the mining company's] proposed mining plan." The bureau's decision is to lift the primary restriction and allow deeper mining, but to require new modifications in the mining plan intended to protect groundwater conditions.

This chapter explains opportunities for public comment in rule-making and in regulating or permitting processes in the implementation of a law. It shows you how to write a formal public comment.

In federal government, after a law is enacted, an executive branch agency begins rule-making to implement the law. Often, this involves developing standards and regulations for administering and enforcing the law. Under the informal rule-making processes of the Administrative Procedures Act as well as other laws and executive orders, the agency must seek public comment on the proposed standards and regulations before they can be put in force. Sometimes, oral hearings are held; typically, written comments are requested.

In state and local governments, public comment might be sought in rule-making, but in many states this is not done as routinely as in the federal government. However, state agencies typically seek public comment in procedures of granting, revoking, or renewing permits for activities that affect public life.

Federal, state, and local agencies welcome any type of comment that can help them make and justify their decisions. The comment might be a technical analysis, a philosophical argument, an opinion based on personal experience, advocacy, or a simple request to hold a public meeting on the proposed action. Responsible agencies will review all written comments. They take seriously the well-prepared comments that suggest realistic and feasible alternatives.

Public comment is important because public policy broadly affects present and future daily life in society and in its environment. A call for public comment invites any member of the public—individuals, communities, organizations—to influence the standards and regulations that affect real lives and livelihoods. As an influential genre, written public comment is underutilized. That's regrettable because public comment is easy to do. Anybody can write a useful comment. The more who do so, the better the likelihood of good government.

Self-governance depends on people's willingness to intervene in the process.

How to Write a Public Comment

Goal: Knowledge of administrative procedures for implementing law, including the public's role in implementation.
Objective: To influence the administration of a law by contributing information relevant to standard setting, rule-making, or permitting.
Product: Formal written comment.
Scope: Pinpointed, specific proposed administrative action.
Strategy: Comments based on your qualifications to respond, whether personal experience, organizational advocacy, vocational or professional background, or specialized knowledge.

While expert commentary is always appropriate, you need not be an expert in order to comment. Administrators want and need to hear from anyone who can make a useful comment. There is no template for public comments, unlike legal briefs. A simple letter can have an impact. If friends or a community organization shares your views, you might want to present a collective comment. You might sign for the group, or all individuals involved might sign, or the organization's officers might sign.

Task #1. Find calls for public comment.

The U.S. government's official source for notifications of proposed rule-making, or standard setting and regulation, is the *Federal Register* published daily. You can find the *Federal Register* either in government information depository libraries or online at http://www.access.gpo.gov/su_docs/aces/aces140.html.

In the *Federal Register*, you will find calls for comment in the section titled "Proposed Rules" or the section titled "Notices." Look for announcements by agencies authorized to act on topics of concern to you.

Alternatively, if you already know the executive branch department, and within it the relevant agency, that administers laws in your area of concern, do not go initially to the *Register*. It can be overwhelming, and you would have to look at the index every day to follow the government's activities on an issue of concern. Instead, first try the website of the relevant department; search there for the relevant agency. If you do not know the relevant agency's name, go to the website of an advocacy group associated with your concern. Browsing there is likely to turn up the name of the relevant department and agency. Then proceed with searching the agency's website for notifications.

If you are concerned about a state issue, you can find calls for public comment in state notifications such as the *Pennsylvania Bulletin* or the *New York State Register*. Every state has one. Familiarize yourself with the index and other finding aids for the state publication you are likely to use often. Alternatively, if you know the jurisdiction for your concern, go first to the website of the state agency with jurisdiction. Or go to the websites of interested associations and advocacy groups to find where you can make a comment on an active issue.

If you want to comment on a local government matter, consult local newspapers. Local government calls for public comment are published in the public notices section of newspapers. Notifications are also posted in local government offices or, possibly, on their websites.

Task #2. Write the public comment document.

In most respects, writing a public comment is like writing any other policy document. The demands for preparation and planning are the same. The same criteria for clarity, credibility, and conciseness apply. One possible difference: Some calls for public input specify the exact information needed. If the call to which you are responding does specify the contents, be sure to provide them as requested. If you have additional information, include it too, but not at the expense of requested contents.

To help ensure that your comment will be taken seriously, include these features and qualities:
- Narrow focus
- Evidence, analyses, and references supporting your view
- Indication of public support of your view
- Positive and feasible alternatives

Before you write, use the method in chapter 2 to plan. After you write, check the product against the general standard (see checklists, chapter 2).

Three Examples

Example 1.   This written comment is a technical analysis of a proposed change in motor vehicle safety standards. The comment was written by professionals in the field of automotive safety and submitted in the rule-making process described in Scenario 1 at the beginning of this chapter.

COMMENT TO THE DOCKET CONCERNING AMENDMENTS TO FMVSS 208, OCCUPANT CRASH PROTECTION

Summary of Comments

Federal motor vehicle safety standards (FMVSS) must, by law, meet the need for motor vehicle safety. This proposal (Docket No. NHTSA 98-4405; Notice 1) purports to meet that need by requiring advanced air bags. In fact, it is primarily written to address the problem of inflation induced injuries and would provide little additional protection.

The worst of the inflation induced injuries resulted in several hundred fatalities to children and out-of-position adults (including those sitting too close to the steering wheel) and from late, low-speed crash air bag deployments. NHTSA [National Highway Traffic Safety Administration] had assumed that manufacturers would conduct comprehensive air bag testing to ensure that inflation would not inflict injury under reasonable foreseeable conditions. It is arguable (although probably not practical policy) that NHTSA could address inflation induced injuries under safety defect provisions of the National Traffic and Motor Vehicle Safety Act [detail omitted].

A key part of this notice proposes two options: (1) tests of air bag systems with dummies in close proximity to ensure that inflation induced injuries are unlikely, or (2) requirements for occupant sensors to ensure that air bags will not inflate if an occupant is in a position where he or she is at risk of injury from the inflating air bag.

In response to the proposed alternatives, we expect manufacturers to choose occupant sensors to prevent air bag inflation for certain occupant situations. This untested sensor technology might actually increase casualties because of inaccurate determinations of occupant risks and degraded reliability from the added complexity.

Experts in the field have suggested a number of potential air bag design and performance features that might reduce inflation induced injuries. The Canadian government and NHTSA deserve credit for their research and analysis in this field despite NHTSA's belated recognition that an official response was necessary. It is not clear which approach would be most effective, or even most cost-effective, but we think it is unlikely that NHTSA's proposed regulation will yield an optimal result.

This notice [of proposal amendments] also fails to address occupant protection challenges involving one to two orders of magnitude more casualties for which feasible technologies are available. These include raising safety belt use to near universality, protection of occupants in rollover crashes, and addressing compatibility problems between passenger cars and light trucks.

Many of these deficiencies can be overcome with a third alternative that retains the simplicity of the original automatic occupant crash protection standard; does not introduce complex, untested occupant sensors; and meets other needs for motor vehicle safety. It depends fundamentally on NHTSA's willingness to propose acceptable, effective inducements for using safety belts.

Problems with NHTSA's Proposal
[Detail not shown here.]

"Please Don't Eat the Daisies"
[An argument that the government should not have to tell manufacturers everything they should or should not do to protect people. Not shown here.]

A Third Option Would Encourage Belt Use
We are proposing that a third option be added to NHTSA's notice that would ensure safety belt use with acceptable and effective belt use inducements built into the vehicle [detail omitted].

NHTSA must recognize that the fundamental problem with its occupant restraint policy is that a substantial minority of motorists do not use safety belts. In fact, a much larger proportion of those most likely to be involved in serious crashes drive unbelted. Nearly universal belt use is critical to any rational occupant crash protection program.

Using the Marketplace
[An argument for useful safety consumer information. Not shown here.]

Diagnosing Problems and Evaluating Change
[Detail not shown here.]

Detailed Comments

Background and Policy
[Detail not shown here.]

An Alternative to the Proposed Amendment
Our specific proposal is that NHTSA add a third option to its notice on advanced air bags. Under this option:

- A manufacturer must install an effective, but not onerous safety belt use inducement in a new motor vehicle of a type that would be permitted under the 'interlock' amendment (15 U.S.C. 1410b) to the National Traffic and Motor Vehicle Safety Act [detail omitted].
- A motor vehicle must meet comparative injury criteria of FMVSS 208 and in addition [in crashworthiness tests using dummies] there can be no contact between the head of the driver or passenger dummy and any part of the vehicle (other than the air bag or belt restraint system) or any other part of the dummy, in a frontal barrier crash at a speed of up to 35 mph with belted occupants [detail omitted].
- Air bags may not deploy under any frontal crash speed barrier impacts below 16 mph [detail omitted].
- Vehicles would be subject to an offset barrier test similar to that proposed in this notice [detail omitted].

Consumer Information and the Market for Safety
[Detail not shown here.]

Discussion

Our alternative would provide occupant crash protection that is at least equal in all respects to that provided by the present standard and NCAP consumer information program [detail omitted].

This proposal would substantially increase belt use and, because of the head impact requirements, would ensure that air bags provide good head protection. Air bags that can meet this criterion would provide some frontal crash protection to the small number of unbelted occupants (who would, of course, be unbelted by their own conscious choice).

If manufacturers would choose our alternative, it would save a minimum of 7,000 lives per year compared with the present FMVSS208, making it one of the most cost-effective standards ever.

(The full comment can be found in the Department of Transportation's Docket Management System [http://dms.dot.gov] by searching for docket number 4405.)

Example 2. This is a technical analysis of a mining permit application. Prepared by a local environmental conservation group, the comment is written by the group's attorney in response to a call for public input on the permitting process described in Scenario 2 at the beginning of this chapter.

TECHNICAL ANALYSIS

July 29, 2001

Michael W. Smith
District Mining Manager
Pennsylvania Department of Environmental Protection
P.O. Box 209
Hawk Run, PA 16840-0209

VIA Hand Delivery

Re: Con-Stone, Inc's, June 7, 2001, application to revise permit #14920301

Dear Mr. Smith:

The Penns Valley Conservation Association (PVCA) has reviewed Con-Stone, Inc.'s, June 7, 2001, application to revise permit

#14920301 to allow removal of the Valentine Limestone below the 1080' elevation. PVCA wishes to work cooperatively with Con-Stone and the Pennsylvania Department of Environmental Protection (DEP) to ensure that mining operations protect the watershed surrounding the Aaronsburg Operation, including Elk and Pine Creeks [state-designated Exceptional Value (EV) stream] and Penns Creek [state-designated High-Quality (HQ) stream]. In that spirit of cooperation, and for protection of those streams, PVCA requests denial of Con-Stone's current application for the following reasons.

1. PVCA recommends retaining special conditions 1, 2, and 4 in Part B, Noncoal Surface Mining Permit No 14920301, Revised July 13, 1999, Special Conditions or Requirements. As District Mining Manager Michael W. Smith said in an August 27, 1999, letter to Con-Stone, "The mining limit of 1080 feet was originally established to keep mining activity out of the average seasonal low water table to minimize the potential for impacts to groundwater and to Spring S-26. We are not convinced that mining below 1,080 feet can be accomplished without added risk of water impact [detail omitted]."

2. To manage the risks of mining below the water table, the proposed amendment calls for phased mining with a progressively deeper penetration of the water table. However, there is a total lack of detail in the permit amendment regarding the specific steps to be taken in the phased mining process. There should be clear language in the permit that stipulates consecutive mining and reclamation and attaches some time schedule and methodology for data analysis and reporting prior to advancing to the next phase of mining.

3. In PVCA's original discussions with DEP, Mike Smith indicated that Con-Stone would have to develop a new infiltration basin system to dispose of the groundwater pumped from the quarry. The permit amendment is contrary to this position as it utilizes the infiltration galleries currently designated for storm water disposal. PVCA requests that separate infiltration systems covered by separate NPDES permits be developed for the storm water and groundwater pumped from the pit.

4. The materials contained in the permit amendment do not adequately describe the hydrogeologic conditions. An appropri-

ately scaled map showing the current pit location, the water table configuration, the location of all boreholes, the sedimentation basin, and infiltration galleries should be prepared. Without water table contour mapping, it is impossible to address issues such as recirculation of the water pumped from the mine [detail omitted].

5. The May 7–10, 2001, pit pumping test performed by the mine operator provides little useful information regarding the extent of [potential loss of water supply] due to mining operations [detail omitted]. If DEP is going to grant the requested amendment, PVCA requests a special condition that Con-Stone is responsible for replacing the water supply for any water losses that result from mining operations.

6. The McWhorter Model used to estimate the volume of groundwater to be intercepted during mining is extremely simplistic. It does not reflect a state of the art effort. In the context of the EV protection of the watershed, a much more comprehensive modeling effort is warranted. "Ideal aquifer" calculations such as those used to calculate inflow to the pit are not applicable in this setting [detail omitted].

7. The permit amendment submission does not address the continuous turbidity monitors currently maintained at Spring 26 and Pine Creek just upstream of Spring 26. The continuous turbidity monitoring should be continued, the costs associated with the monitoring should be borne by Con-Stone, and the data collection system discussed at the December 15, 2000, meeting with DEP should be implemented [detail omitted]. In addition, monitoring should be expanded, again at Con-Stone's cost, to be more complete by including all biological and chemical monitoring required by DEP's water quality anti-degradation regulations and implementation guidelines.

8. The mining permit amendment does not address the issue of the potential for fines (fine-grained debris) contained in the backfill to be mobilized into the groundwater system during surface water runoff events or during high water table conditions [detail omitted]. A detailed assessment of the potential for significant contamination of conduit flow system from the fines contained within the backfill should be performed.

9. The mining permit amendment says that "if continuous flow rates greater than 475 GPM are experienced, then further extractions within that area below the water table will not occur unless drought conditions result in an additional decline of the groundwater table." This statement should be revised to specify what "additional decline" is necessary to allow mining operations to resume.

10. The permit amendment submission does not appear to be signed and sealed by a licensed professional geologist.

11. PVCA believes Con-Stone must apply for a new or revised NPDES [National Pollution Discharge] permit for the proposed quarry dewatering activities.

12. If DEP is going to allow any mining below 1080 feet, PVCA requests that the special conditions 3 and 5-21 in Part B, Non-coal Surface Mining Permit No. 14920301, Revised July 13, 1999, Special Conditions or Requirements continue to apply to all mining operations at the Aaronsburg Operation.

13. PVCA does not believe that Con-Stone has complied with all necessary pre-permit requirements under DEP's water quality anti-degradation regulations and implementation guidelines [detail omitted]. Further, the application has not sought review by local and county governments to ensure compatibility with applicable regulations, ordinances, and comprehensive plans and to allow government to identify local and regional environmental and economic issues that should be considered.

Sincerely,

J. Thomas Doman
Chair, Watershed Committee
Member, Board of Directors, PVCA

cc: Jeff Confer, Con-Stone, Inc.
     Hon. Jake Corman, Pennsylvania Senate
     Hon. Kerry Benninghoff, Pennsylvania House of Representatives
     Pennsylvania Trout Unlimited

Example 3.   This is a letter by a citizens group requesting the public meeting described in Scenario 2 at the beginning of this chapter.

CITIZENS GROUP'S LETTER

July 16, 2001

Re: Application for amendment for Con-Stone mining permit #14920301, Aaronsburg Operation

District Mining Manager
Department of Environmental Protection
Bureau of Mining
Hawk Run District Office
PO Box 209
Hawk Run, PA 16840-0209

On behalf of the Aaronsburg Civic Club I am requesting a public conference on the proposed amendment to the above permit. Again this year, we are offering our facility, the Aaronsburg Civic Club Community Building, for that purpose. As you are aware last year's public meeting was well attended and provided an opportunity for residents to state their concerns and for Con-Stone and DEP to address them. This is as it should be in a free and democratic society.

I strongly urge you to hold a public meeting on the latest proposed permit revisions. Two concerns that have been brought to my attention are (1) the potential degradation of underground and surface water, and (2) mining on land previously designated for storage.

Please contact me to reserve our facility.

Sincerely yours,
Earl Weaver, President
Aaronsburg Civic Club

What These Examples Show. All three examples exhibit qualities of effective public policy communication (see checklists, chapter 2). They are narrowly focused, provide supporting evidence or analysis, and refer to public support. Examples 1 and 2 offer clear alternatives to the proposed action (see Task #2, this chapter).

Example 1 on auto safety and Example 2 on mining show public support of the viewpoint, not by using references to opinion polls or

other statistical measures but by invoking public authority. Those examples invoke federal motor vehicle safety standards and state environmental protection regulations to show public endorsement of
their position. Example 3 invokes the authority of "sunshine" man-
dates for public access in support of a local civic group's wish to participate in a government decision affecting their town.

Examples 1 and 2 are detailed and technical. Although the contents are organized and subheadings are provided to aid comprehension, some of the details might be moved to an appendix. However,
the choice to use that option should depend on the writer's knowledge of the circumstances in which the documents will be read and
used. As noted in the commentary on earlier examples, writers
should be certain that all readers will have access to the entire document before deciding to move crucial details to an appendix.

Taken together, the three examples show the variety of uses for
the genre of public comment. Technical experts and concerned citizens alike can use this genre effectively to intervene in the administration of policy.

CONCLUSION: ❖ ❖ READY FOR CHANGE

Framework, Function, and Form

This book gives you a framework for understanding the public policy making process and the functions of communication in the process. It introduces you to forms of communication in current use.

Change in the process is inevitable, and communication will change accordingly. The best illustrations of change come from transnational public policy making. For instance, when trade agreements between nations make partner nations' environmental protection regulations or worker health and safety laws unenforceable, structural reform is required. Communication will alter accordingly.

In two recent examples, the North American Free Trade Agreement (implemented in the 1990s by the United States, Canada, and Mexico) and the European Union (initiated in the 1980s by former independent nations of Europe), structural reorganization has led to the creation of genres such as directives for communicating transnationally. A European Union directive, for instance, will affect the economies, societies, and environments of more than twenty nations. To implement the policy communicated by a directive, translations of language, political concept, governmental procedure, and communication practice are involved.

Domestically, ordinary political change will continue to drive adaptation, too. When, for instance, the political majority in a national government changes, major alterations in policy, procedure, and practice typically result.

Thus, you might someday want or be asked to find, adapt, or create new communication forms to respond to change. In any case, the framework presented here can be helpful. It recognizes that institutional policy making depends on a sequence of actions—from defining a problem to intervening in administration of policy. The framework also reminds you that democratic action depends on communication, and it prompts you to consider the context for communicating. Thus prepared, you are ready for change. You can respond with control and creativity.

❖ Index ❖